PLANT PARTNERS

PLANT PARTNERS

SCIENCE-BASED COMPANION PLANTING STRATEGIES *for the* VEGETABLE GARDEN

JESSICA WALLISER

Storey Publishing

EDITED BY Carleen Madigan and Hannah Fries
ART DIRECTION AND BOOK DESIGN BY Michaela Jebb
TEXT PRODUCTION BY Jennifer Jepson Smith
INDEXED BY Christine R. Lindemer, Boston Road Communications

COVER PHOTOGRAPHY BY © Angelo Merendino Photography, LLC, back t. & b.c.; © Derek Trimble, front b., back b.c.l.; © Gillian Pullinger/Alamy Stock Photo, front t.; © Saxon Holt, back b.c.r.; © sever180/stock.adobe.com, back b.l.; © Tim Gainey/Alamy Stock Photo, back b.r.

INTERIOR PHOTOGRAPHY BY © Derek Trimble

ADDITIONAL INTERIOR PHOTOGRAPHY BY © Angelo Merendino Photography, LLC, v, viii, 4, 7, 26, 43, 59 r., 85, 92, 100, 106, 115–117, 121, 144, 156, 166, 177, 184

Additional photo credits appear on page 205

Storey Publishing
210 MASS MoCA Way
North Adams, MA 01247
storey.com

Printed in China through World Print
10 9 8 7 6 5 4 3 2 1

Library of Congress Cataloging-in-Publication Data on file

To Niki Jabbour and Tara Nolan,
my amazing SavvyGardening.com
partners and soil sisters.

CONTENTS

FOREWORD

For years gardening "experts" told gardeners to place particular plants next to each other to control insects or disease, or just because those plants "work well" together. Unfortunately, there really wasn't much scientific data to support their recommendations, so, despite good intentions, these combinations were mostly just someone's best guess. Today, scientists still don't have all the answers, but recent studies have produced greater insight into the interactions between plants, providing clarity about how they might benefit one another.

With this book, Jessica Walliser has written something for the thoughtful gardener: the one who wants to understand exactly what they're doing and why, and who is willing to observe, think through, and experiment with their garden choices. The bibliography at the back of this book lets us know that it's the product of real research. This research, along with the experience and careful observations of a committed horticulturist, makes *Plant Partners* a valuable gardening reference.

If you want garden dogma, then go somewhere else. This is a reference for the open-minded gardener who is willing to think through their gardening choices. It gives us the science-based information we need to make informed decisions when choosing which plants to place next to one another. *Plant Partners* offers more than just specific plant pairings; it encourages us to think about the relationships between plants, so that we can grow our best garden ever.

— Jeff Gillman, PhD

Director of the UNC Charlotte Botanical Gardens and best-selling author of *The Truth about Garden Remedies* and *Decoding Gardening Advice*

As a horticulturist, garden educator, radio host, and self-described science nerd, I've received hundreds of questions and comments from people about the merits of companion planting. They wonder if their beets will grow better if they plant them next to their beans. Or if planting onions next to broccoli yields a better crop of both. I've also had people swear that when they plant dill next to tomatoes, they don't get hornworms. Or that when they plant carrots next to basil, both of the plants thrive because they "like" each other. I've often struggled to provide responses to those inquiries because I haven't had the science to back them up.**

You see, while companion planting has a long history among gardeners, it's a history filled with folklore and conjecture, often at the expense of sound science. But times are changing. New and existing research from universities and agricultural facilities around the world isn't necessarily validating the long-held companion planting techniques and beliefs that have been around for generations. Instead, it's pointing us toward a whole new way to companion plant. A way that approaches the garden as an ecosystem comprising many different complex layers of plants, fungi, and animals, all of which are connected in a massive web of life.

Modern companion planting isn't about what plant "loves" growing next to what other plant. It's about using plant partnerships to improve the overall ecosystem of the garden and create a well-balanced environment in which all organisms thrive, from the tiniest soil microbe to the tallest corn plant. It's about pairing plants such that one plant provides a benefit to the other in terms of an ecosystem service. For example, maybe plant A provides food for plant B. Or maybe plant A controls weeds around plant B. Or maybe plant A attracts beneficial insects that control pests on plant B. As you'll soon learn, plants can benefit each other in these and many other ways.

For generations, companion planting has been defined as the close pairing of two or more plant species for the purpose of enhancing growth and production, or trapping or deterring pests. It's a definition that still rings true, even as companion planting undergoes a much-needed science-based reboot. Farmers and gardeners now have myriad credible studies, controlled experiments, and fact-based research to rely on. Some of these studies provide a pretty jaw-dropping look at mutualism within plant communities, between plants and animals, and even between plants and fungi for the betterment of one or more of the organisms involved. It's all a big web of connections out there in the garden, and it's time we started paying attention to those connections and how they can make us better gardeners.

In the pages of this book, you'll find both broad and narrow looks at many of the studies I mention above. My aim, however, is to put a modern twist on the practice of companion planting by offering gardeners ways to use these findings in their own landscapes to improve plant health, yields, and productivity.

Chapter 1 explores many of the possible benefits of modern companion planting and takes a deeper look at why this long-standing practice needs to be approached in a whole new way. We'll examine the role polycultures, plant associations, and interplanting play in the garden and look at why so many scientists dislike the term *companion planting*. One long-held yet scientifically unproved companion planting theory is that plants whose extracts have a similar crystalline structure grow well together. Other outdated theories are based on the growth patterns or "energies" of a plant. We now know, however, that things like fungal associations, resource competition, chemical messaging, plant diversity, and nutrient absorption have far greater impacts on how well one plant grows alongside another than the myth-based theories could ever tell us. This chapter

takes a look at each of these influencing factors and how they translate into a successful companion planting strategy.

Chapters 2 through 8 each tackle a common gardening problem by using appropriate plant partnerships to overcome it. You'll find a chapter on companion planting to boost soil health, one on controlling weeds, another on using plant associations to limit pest damage, and much more. Whether you want to manage plant diseases, improve pollination, or encourage a naturally high population of pest-eating beneficial insects, these chapters offer everything you'll need in the form of one or more methods of companion planting. Each chapter begins with an introduction to the problem, followed by several research-backed plant combos and other strategies that use companion planting as an effective tool to solve that particular problem.

Every chapter in this book speaks to how important an ecosystems-based approach is to the modern landscape. Even as many of our yards and gardens — along with the fields, forests, deserts, and prairies of this Earth — have become less and less diverse over the years, our understanding of the value of environmental diversity has never been greater. When you implement some of these concepts and plant partnerships in your home garden, the impacts of your actions will extend well beyond your garden's gate.

As you'll come to see, today's companion planting methods generate real and measurable benefits. *Plant Partners* aims to redefine the science of successful plant partnerships by examining the many ways one plant influences another. After closing the final pages of this book, I hope that you, like me, have a whole new appreciation for companion planting in its modern form.

THE POWER OF PLANT PARTNERSHIPS

How Does Modern Companion Planting Work?

One might assume a garden is made up of independent parts, each managed independently by the gardener. But in truth, the many parts of a garden are in constant interaction, much like in a wild ecosystem. Yes, the gardener plays a role in managing these interactions, but all of the different components — from the plants and soil to the insects, fungi, and microorganisms — have ongoing effects on each other whether managed or unmanaged.

Modern companion planting strategies recognize these interactions and acknowledge gardens as ecological habitats in which plants are capable of influencing each other in a variety of ways, including through the chemical signals they produce, the underground network of fungi in and around their roots, and the toxic compounds some of them exude. Many strategies are aimed at influencing the insect world. Though in many ways we're only just beginning to understand the complexities of these interactions, we can already use our current knowledge to become better gardeners and build healthier, more productive gardens.

A MODERN TAKE *on* GARDENING

Today's ornamental gardens are, in many ways, different from those built by past generations. For decades, plantings of perennials, annuals, trees, and shrubs were almost exclusively focused on the ornamental value of the garden. Now this approach is shifting to one that is much less centered on human aesthetics and more focused on providing wildlife habitat and conserving resources. While a huge chunk of the "gardening public" certainly still thinks of gardens exclusively from a human point of view (What color is that flower? How tall does that shrub grow? How hard is that plant to maintain?), there are now more gardeners than ever who create and cherish gardens for their ability to provide something beyond their appeal to the human eye. These gardeners plant flowers, trees, shrubs, and ground covers to attract and support pollinators, to absorb rainwater runoff, to sequester carbon, to conserve irrigation water, to filter pollutants from the air, to create habitat for a variety of animals and insects, to build healthy soil, and to improve the overall biodiversity of their little part of the world. Yes, it's a bold jump to make, but millions of gardeners are now looking at their favorite hobby as an opportunity to make a positive difference in the health of the planet.

Despite all of these changes in the world of ornamental gardening, times don't seem to have changed much in the vegetable patch. The victory gardens planted a few generations ago mimicked farm-style plantings with their highly organized "soldiers in a row" design of row after row of single crops. Unfortunately, the vast majority of modern vegetable gardeners here in North America still view the garden as a place of order, where crops are planted in straight rows to become a mini monoculture of sorts. Perfect lines of vegetables are flanked on either side by empty rows of bare or mulched soil, and flowering plants that do not produce edible crops are relegated to the garden's edge or are not included at all. Vegetable gardens have been slow to evolve. For some reason, many vegetable gardeners don't view their garden as a working ecosystem, seeing it only as a space for maximizing yields for human consumption.

However, even a slight shift in thinking can result in big changes in the vegetable plot. Since food gardens are the primary places where companion planting strategies come into play, gardeners who focus on creating mixed plantings that include a diversity of vegetables, fruits, herbs, and flowers all growing together are benefiting not just themselves but also the garden ecosystem as a whole. They're supporting a diversity of pollinators, staving off soil nutrient depletion, naturally deterring pests, and more — all while still maximizing their yields for human consumption, just in a different way. Raised beds, containers, planter boxes, and in-ground plots filled with a combination of plants are far from a monoculture, and as you'll come to learn, when those combinations are made with specific plant partnerships in mind, the benefits are magnified.

Whether vegetable or ornamental, gardens are ecosystems capable of supporting biodiversity, filtering rainwater, sequestering carbon, and providing many other services.

THE BENEFITS *of* COMPANION PLANTING

Let's spend a little time discussing some of the possible benefits of companion planting and look at how these mixed-planting strategies can affect vegetable gardens. Most of the benefits that companion planting strategies provide fall into one of the following seven categories.

1 **Reduced pest pressure.** Minimizing pest damage tends to be the most sought-after benefit of companion planting. The research into it is mind boggling, with countless studies looking at everything from how pests find their host plants to strategies for luring pests away from desired crops before they can cause significant harm. Companion planting to reduce pest pressure utilizes luring, trapping, tricking, and deterring pests to keep vegetable garden damage at a minimum.

2 **Reduced weed pressure.** A reduction in weeds without the use of herbicides is another possible benefit of some companion planting strategies. You'll be introduced to the science of allelopathy in chapter 3 and how it can be used to impact weed growth in the garden. Companion plants can also serve as a living mulch to reduce weed pressure through crowding and shading.

3 **Reduced disease pressure.** Perhaps surprisingly, companion planting is being studied for its ability to suppress certain plant diseases. Though this branch of companion planting does not appear to be as well studied as some others, the interplay between disease organisms and the plants they affect can indeed be influenced by certain companion planting strategies.

One possible benefit of companion planting is improved pollination.

4 **Improved soil fertility or structure.** Green manures and cover crops have long been used as companions to vegetable and grain crops, though primarily in larger agricultural operations. Home gardeners, too, can reap the benefits of these soil-building strategies when used properly, even on a small scale. Soil structure can also be improved by using certain plant partnerships, including those aimed at breaking up heavy clay soils or improving

These parasitic wasp larvae emerging from a tobacco hornworm are a classic example of biological control in the vegetable garden.

the condition of the soil through the presence of root exudates (compounds produced and excreted by the roots of living plants). Other companion planting techniques assist in nitrogen transfer to improve the fertility of the soil.

5 Improved pollination. Companion plants are capable of improving garden yields by increasing the number and diversity of pollinators in the area. By carefully selecting plant partnerships that encourage and support the specific species of bees known to pollinate target crops, pollination rates may be improved.

6 Improved biological control. Another benefit of certain companion planting techniques is an increase in the population and diversity of the many beneficial insect species that dine on common garden pests or use them to house and feed their developing young. Partnering plants that attract and support pest-eating insects results in greater biological control and fewer pest outbreaks in the garden. Companion plants can create habitat for these "good" bugs, as well as provide the insects with essential nutrition in the form of pollen and nectar. Plant partners can also be chosen for their ability to serve as "banker" plants that are intentionally grown to attract and support pests so that beneficial insects can use them as a food source when pest populations are low in crop plants. This practice may help improve or equalize the seasonal population of beneficial insects by giving them a reason to stick around. There are many studies under this umbrella that reveal some wonderful ways for gardeners to encourage a healthy balance between pests and the predatory and parasitic insects that help control them.

7 **Improved aesthetics.** While monocultures do occasionally deserve a place in the garden (think slope-covering ground covers, for example), you'd be hard pressed to find a garden visitor who doesn't find a mixed planting more attractive than a monoculture. Unlike on farms where row planting is necessary to simplify mechanical harvesting, home gardens are the perfect places to feature a mixed-planting design. And since layered gardens with many levels of plant structures and many growth habits, from ground covers to trees and everything in between, are more inviting to a broader diversity of insects and other wildlife, the aesthetics of companion planting come with additional benefits.

Six of the seven benefits just discussed are featured in their own subsequent chapters, each approaching companion planting from a problem-solving point of view. Since garden aesthetics are both personal and opinion based, I've left that topic out of the mix. However, I do introduce one type of companion planting that undoubtedly offers the benefit of improved aesthetics, among others: chapter 4 is all about companion planting for support and structure. It highlights plant partnerships that combine one plant that serves as a living structure or trellis with another that needs something to climb.

How Close Do Companion Plants Need to Be?

An oft-asked question of many gardeners when it comes to companion planting is "How close do the plants have to be for them to be considered companion plants?" The answer is "It depends." Physical proximity as well as planting time varies quite a bit, depending on the desired results and the methodology behind each specific combination. Sometimes the companion plants are planted at the same time, while other times the plant partners are planted in succession. Sometimes the companions are in physical contact, while in other cases the two plants might be several feet apart. Plant partnerships don't necessarily have to occur simultaneously. As you are introduced to the benefits and techniques of the many plant combinations in this book, you will also learn the particulars of timing, physical proximity, and other details.

COMPANION PLANTING *by* OTHER NAMES

I'd bet that most of the scientists who study plant partnerships don't call the subject of their work "companion planting," probably due in no small part to the fact that the term has developed some negative connotations over the years. When applied to growing on a larger scale, terms like *polyculture* and *intercropping* are more often used. Let's take a closer look at these two terms and what they mean.

Polyculture is defined as an agriculture system in which multiple plants are growing in the same space to reflect the diversity of a natural ecosystem and create an environment where pests and diseases don't spread as easily as they do in a monoculture. Companion planting is one way to create a polyculture.

Intercropping is the practice of growing multiple crops in the same field area to promote beneficial results. On the smaller scale of the home garden, it's called interplanting. Intercropping or interplanting creates a polyculture. There are several types of intercropping/interplanting:

+ When crops are mixed such that partner plants are blended together with no distinct rows or other formal arrangement, it's called mixed intercropping/interplanting.
+ When the crops are mixed together in alternating rows, it's called row intercropping/interplanting.
+ On farms, when a second crop is planted right into an existing crop just before harvest of the first crop, it's called relay intercropping.

Even small vegetable gardens like this one can be a polyculture filled with a diversity of plants.

+ In a closely related practice, intercropping can also take the form of different crops being planted at different times of the year. This is called crop rotation.

Regardless of which term is used, it all boils down to the idea of planting two or more plant species in close proximity for some benefit to one or more of them — which is the basic definition of companion planting as used in this book. Companion planting *is* intercropping/interplanting to create a polyculture, but on a smaller scale than what occurs in a commercial farm environment. While it's true that some aspects of polyculture or intercropping may not translate well to a smaller scale, few if any of them will cause harm to the garden or the gardener. One might even say that interplanting is the "new" companion planting.

So when you come across the terms *intercropping* and *polyculture* when looking at research regarding these techniques and their effectiveness, know that the principles of companion planting fall under what is essentially the same umbrella, just on a smaller scale.

HOW TECHNIQUES *Are* TESTED

In the scientific literature, there is a whole host of credible studies exploring mutualistic relationships among multiple plants, and between plants and insects, fungi, and/or microbes. Both fascinating and groundbreaking, these are the studies we must look to when determining the best plant partnerships to use in our backyard gardens. But how and where and why do these studies take place? And how well can we expect the information they yield to translate from the study environment into our backyards? In other words, can we expect comparable results?

Research into the workings of polycultures, intercropping/interplanting, and companion planting techniques takes place in universities, government research facilities, nonprofit agricultural research institutes, and various corporate and private research programs, as well as on working farms via numerous educational and research programs around the world. Research facilities and demonstration farms provide study locations that mimic the growing conditions found on real-life farms. While they are often divided into study plots and sometimes are home to multiple simultaneous studies, these facilities provide the right environment for the careful implementation of research and the collection of data. They cannot, however, perfectly re-create the conditions of a real working farm with its many variables and unpredictable situations.

On the other hand, plenty of studies *are* conducted on working farms, where conditions are as real as you can get. Plant partnerships are studied in the fields even as harvestable crops are being grown, and data are collected and analyzed on-site. Some farms and research facilities set up small-scale field studies that

imitate more of a garden setting than an agricultural one. Nonetheless, constraints and challenges are always present when studying companion planting methods, especially because most of the studies take place in an outdoor environment, where weather, topography, soil conditions, microorganisms, insect populations, and any number of other unpredictable factors may influence the outcome and cause mixed results. Even so, scientists are constantly conducting research that results in actionable, effective ways we can utilize plant partnerships to our benefit.

Size undoubtedly plays an additional role in how the outcome of many studies translates from farm to garden. The important thing to remember is to be flexible and understand that the results of your companion planting efforts may differ from the results of a study that was conducted on a farm-sized plot. Despite all of this, these larger-scale studies are more often than not the best research we have available regarding which companion planting partnerships may work and which likely will not. In seeking out techniques to include in this book, I purposely looked for those that were backed by studies conducted in environments similar in size and scale to home gardens.

Regardless of the size of a study's scope and where it occurred, one point that is across-the-board useful — no matter the size of your growing plot — is the avoidance of monocultures. This is probably the most critical aspect of companion planting. Diversification is key, whether on a home scale or in an agricultural setting.

You, the gardener, play another key role in successful companion planting. Yes, studies provide crucial information, but it's important to evaluate and record your own results. Careful observation and note taking are bound to increase your chances of success. Write down what works in your garden and what doesn't, as well as any changes in the health and/or performance of your plants. While your efforts may not be 100 percent scientific or ever appear in a published study, they are invaluable. The number of possible plant partnerships is countless, and researchers will never be able to carefully observe and study all combinations of plants to evaluate whether or not they prove beneficial. But in your own garden, you can examine and assess different plant partnerships for their ability to achieve your desired results.

Now that you know the benefits of companion planting and how to look at the practice in terms of its ability to create an all-important polyculture, let's look at the many ways plants interact with each other and how we can leverage these interactions to create successful companion planting strategies.

A diverse garden features both vegetables and flowers.

PLANTS AREN'T PASSIVE

Contrary to what you might assume, plants are not passive organisms at the mercy of their environment. Instead, when plants are subjected to any number of stimuli, they respond in ways we humans are just beginning to understand. Think about it for a second: plants are required to grow with limited resources, compete with each other, survive attacks from various herbivores and disease-causing organisms (pathogens), and reproduce, all while staying in a single spot their entire lives. In order to survive, plants have evolved an incredible number of strategies to deal with these challenges.

Most of the strategies plants use to adapt to their environment are ones we humans can't see. Some are defensive in nature, meant to protect the plant from pests and diseases, while others alter the environment around the plant to make it more suitable for themselves or other organisms to grow in (or less suitable, in the case of pests and competitors).

When it comes to defensive strategies, while some plants have evolved to have thorns, hairy leaves, sharp needles, or other visible adaptations, most of the ways a plant defends itself are in the form of chemical compounds produced within the plant itself. These include nicotine, caffeine, tannins, quisqualic acid, strychnine, cardiac glycosides, and quinine, among one hundred thousand others that have been identified thus far. Many plants produce these defensive chemicals all their lives; others produce them in reaction to injury.

In addition, many plant species emit volatile chemicals into the air as an alarm call of sorts — to warn neighboring plants about insect invaders or to attract predatory insects to feed on leaf-munching pests. Indeed, plants have evolved an incredible number of ways to protect themselves, in spite of their inability to flee danger.

Spines are one form of physical defense some plants use to protect themselves.

Plants can alter their environment to improve their own growth and reproduction or to inhibit the growth and reproduction of other organisms living near them. Yes, to be certain, plants are at the mercy of their environment, but they also have clever tricks up their evolutionary sleeves. Some plants grow tall enough to shade other plants, inhibiting or enhancing their growth. Or they may grow large enough to outcompete nearby plants for water and nutrients. Plants can change the soil they're growing in through the production of root exudates. Such alterations modify the chemistry of the soil and the balance of microbes living there, making conditions better or worse for competing species. Some plants inhibit the growth of other species by producing root or plant exudates that are toxic to their neighbors, therefore giving them a competitive edge.

Plants compete with each other for shared resources, including light, nutrition, and moisture.

HOW PLANTS AFFECT EACH OTHER

There are many ways one plant affects another's health and growth. So let's delve more deeply into the five primary ways plants influence each other and discover the valuable role these factors play in forming plant partnerships that can make us better gardeners.

1 The Use of Shared Resources

Plants need certain resources to survive. While some species may require very little of a certain resource, others may demand a lot of it. There are plants that have evolved to have minimal water needs, while others need a constant supply. Some plants call for little in terms of mineral nutrition; others are heavy feeders. Regardless of the level of resources each species needs, plants influence each other by constantly competing for the same pool of available resources. However, they can also influence each other by sharing resources. Let's look at both sides of this coin.

RESOURCE COMPETITION

Competition refers to a relationship that has negative effects caused by a reduction of available resources, in this case due to the presence of neighboring plants. It's a very important factor in plant health. The effects of competition can be easily observed (weeds overrunning your garden, shading crops, and pulling nutrients from the soil, for example), or they can be subtle (the slowed growth of a tree due to shade from a highly competitive canopy of other trees, for instance). Competition for resources occurs in both managed garden environments and natural habitats. It is a driving factor in plant growth, reproduction, the structure of plant communities, and even evolution.

Weeds create stress for both plants and gardeners.

Plants compete for three primary resources: water, nutrients, and light. (Though they need CO_2 to survive, there's plenty of that in our atmosphere, making CO_2 competition a nonissue.) Because each of these three resources is highly complex, the ways that plants compete for them are diverse and complex, too. Plants may compete for nutrients by maximizing their root growth to acquire nutrients from the soil before neighboring plants do. They may extend their canopy higher to optimize their rate of photosynthesis and shade out their neighbors, in turn reducing the growth of the plants around them.

Competition generates stress, which can influence plant growth in many ways. All environments have a limited supply of necessary resources, and when two or more individuals are competing for those same resources, they use whatever means available to acquire them. Plants in resource-poor environments evolve ways to overcome some competitive challenges. Think of cacti, with their fleshy, succulent tissues capable of storing water for a long period of time, or the ability of certain alpine plants to thrive in lean, rocky soils with minimal nutrition.

As you'll come to see, competition plays a role in some of the most effective methods of companion planting, particularly those presented in chapter 3.

RESOURCE SHARING

In addition to competing for available resources, plants may sometimes share them. While light is a difficult resource to share, some plants are known to limit or maximize the growth of their canopy as necessary to create an ideal microclimate for young seedlings to thrive. Tall trees may compete with the plants growing beneath them for light and nutrients, but in some cases, they may also draw water from the deeper reaches of the soil and pull it closer to the surface where more shallow-rooted plants can access it. Plants can redistribute water and nutrients within the soil as they grow and develop, sharing these resources with other plants in their vicinity.

There's good evidence that plants may even be able to recognize their kin. In studies, roots of plants grown from seed collected from the same mother plant and grown together in a pot were not as aggressively competitive as the roots of plants grown from seed collected from different mother plants. However, the question remains as to whether the nonaggressive roots were due to "friendliness" among kin, or the more aggressive roots were due to increased competition among strangers.

Resource sharing also takes place via an intricate underground network of fungal associations, which is discussed in item number 4 on this list.

While trees do compete with smaller plants and trees growing beneath them, in some cases, they may actually share resources.

2 Improved Nutrient Availability and Absorption

Some plants are capable of influencing the growth of other plants by increasing the amount of nutrients available to them, as well as improving the absorption rates of those nutrients.

Plants acquire nutrients in several ways. They may absorb the nutrient with water (as with calcium and magnesium), they may uptake the nutrient when the root bumps into it as it grows (as with most nutrients), or the nutrient may come into the root via diffusion, which occurs when the concentration of the nutrient inside the root is less than the concentration of it in the soil surrounding the root (as occurs primarily with potassium and phosphorus).

The movement of nutrients within the soil itself is dependent on many factors, including the soil's structure, the concentration of nutrients present in the soil, how strongly the nutrients are bound to the soil, and how mobile that particular nutrient is within the soil. One plant can improve the ability of another plant to absorb nutrients by altering one or more of these factors. Some plants are capable of improving soil structure by increasing the amount of organic matter in it or simply through the penetrative action of their roots. Other plants can increase the availability and accessibility of certain nutrients within the soil through processes like nitrogen fixation and nutrient "pulling." And some plants can increase the amount of certain nutrients found within the soil via the organic matter they leave behind after they decompose. Plants that perform these kinds of functions in a farm or garden setting are called green manures. Green manures are used to positively influence nutrient availability and absorption, making them valuable companion plants.

Green manures leave behind organic matter after they decompose, improving the quality of garden soil.

When it comes to green manures, there are none more notable than those that belong to the legume family. Nitrogen-fixing plant species in the legume family convert nitrogen from the air into a form that's available to other plants (more on this process in chapter 2). Legumes provide this essential nutrient to neighboring plants and microbes, as well as to crops planted in the same area at a later time. While you may think green manures are useful only for large farms, they can have a big impact on small-scale growing, too.

3 Chemical Messaging Systems

The communication system plants and insects use comes in the form of chemical signals, called semiochemicals, that are released into the air or soil. Some of these signals serve as a means of communicating within the same species (think of the pheromones an insect emits when in search of a mate), while others are capable of crossing kingdoms and enabling plants to message insects.

When it comes to communication between plants, these semiochemical messages come in many forms. Pest-infested plants may emit semiochemicals known as herbivore-induced plant volatiles (HIPV), or green leaf volatiles, into the air. Depending on the exact chemical composition, the message may be a warning to neighboring plants, alerting them that it's time to steel their defenses and begin to produce the defensive chemicals I mentioned earlier. But these HIPVs can also lure in the particular species of beneficial insect most likely to prey upon the specific pest present on the plant. These compounds, which can travel anywhere from a few inches to hundreds of yards from their source, are detected by the predator and used to locate its prey. In essence, the plant is messaging predators to come to its aid.

Plants send chemical messages into the soil through their roots.

There are many other ways plants communicate with each other via chemical messages to influence the growth of their neighbors. Studies have found that some plants living in crowded conditions excrete chemical messages into the soil that cue neighboring plants to grow more quickly to avoid being outcompeted, while others may exude chemical messages to inhibit the growth of their neighbors. In forests, when trees' branches touch, they alter their growth either to restrict their size to avoid contact and competition (a phenomenon called canopy shyness) or to grow more aggressively in an attempt to outcompete their neighbors. These responses appear to be driven not just by physical contact in the trees' canopies but also by root secretions in the soil that send chemical messages from plant to plant.

Chemical messaging also takes place via the underground network of threadlike fungi that connects the roots of one plant to another. Which leads us to the next way plants affect one another . . .

Speaking Their Language

Scientists have learned to mimic plants' chemical messaging system by developing artificial semiochemicals. Synthetic herbivore-induced plant volatiles (HIPVs) are already being used by some commercial growers and even home gardeners to draw in predatory insects when a pest infestation is occurring. This is currently done through the use of controlled-release dispensers placed throughout the field or garden, but it's a complicated affair. Evidence shows that these artificial attractants do, in fact, lure in beneficial insects, but the jury is still out as to whether these good bugs stick around and whether they actually eat more pests while they're present.

4 Fungal Associations

We've known for several decades that plants communicate with each other (and the insect world) through volatile chemical signals they release into the air and soil. We now know that plants communicate belowground in another way: Various species of fungi known as mycorrhizae colonize the roots of almost every plant on the planet. Some plants cannot live unless a specific mycorrhizal fungus has colonized its roots. Mycorrhizal fungi receive carbohydrates from their plant hosts and transfer nutrients from the soil into the plants and even from one plant to another. They also help strengthen the plants' resistance to drought stress by increasing their ability to absorb water for plants in times of drought, and there's evidence they may even influence visits by pollinators. This network of threadlike fungal roots (hyphae) extends far beyond the actual root system of the plant,

Mycorrhizal fungi growing in plants' root systems create a vast network through which plants can share resources and send out alarms.

allowing the fungi to bring in soil nutrients that would otherwise be inaccessible to the plant.

New studies point out that the mycorrhizal network can aid in plant communications in addition to transferring nutrients. Researchers found that the fungi acted like a telephone network between plants, passing an alarm signal regarding an aphid attack from an aphid-infested plant to uninfested ones, leading the uninfested plants to strengthen their chemical defenses.

To add yet another layer of complexity, some plants may influence exactly which species of fungi colonize their roots by altering the composition of the root exudates they produce. The species of fungi present within the soil can then go on to influence which other plant species are able to grow nearby. It's an incredibly complex network of connections between plants and fungi that has amazing impacts on plant growth.

5 Allelopathy

Allelopathy is the ability of one plant to produce and release chemicals that inhibit the growth of other plants. Essentially, it's a form of chemical competition. Known as allelochemicals, these substances can be found in any part of a plant, including the leaves, stems, flowers, roots, and even fruits. They can also be found in the soil beneath and around an allelopathic plant. These chemicals affect surrounding plants in many ways. They may inhibit root or stem growth, restrict nutrient uptake, or stifle mycorrhizal relationships.

Because different plant species have different sensitivities to particular allelochemicals, some plants might show no signs of negative impacts when growing near an allelopathic plant while other plant species may wilt, drop leaves, or even die. Allelochemicals persist in the soil, too, sometimes long after the plant that produced them dies or is removed. This can affect subsequent plantings and neighboring plants for some time, depending on how long it takes for the specific allelochemicals to biodegrade.

There are many examples of allelopathic plants in all types of ecosystems. The common invasive weed known as garlic mustard (*Alliaria petiolata*) is an allelopathic plant. This weed secretes an allelochemical from its roots that doesn't directly inhibit the growth of other plants but instead inhibits the growth of the mycorrhizal fungi that are known to support a diversity of trees, thereby affecting our forests in ways beyond what was previously thought. The invasive tree known as tree of heaven (*Ailanthus altissima*) is also allelopathic, a trait that helps it outcompete many other plant species living nearby. And Canada goldenrod (*Solidago canadensis*) produces allelochemicals that inhibit the growth of plants found living around it.

While you may perceive allelopathy as a big negative to plant growth, it can be harnessed for surprisingly positive results. In natural plant communities, plants capable of producing these chemicals have a competitive advantage over their neighbors. But in farm and garden environments, the power of allelopathy is used to limit weed growth. Winter rye grass (*Secale cereale*) is one of the most commonly studied and utilized examples of an allelopathic plant. There are some 16 different allelochemicals found in winter rye that prevent weed seed germination but do not harm transplants of peppers, tomatoes, eggplants, and other vegetable crops that are grown in rye residue. Because of this, rye is well suited to companion planting techniques aimed at weed control, as you'll discover in chapter 3. Other common cultivated plants with allelopathic effects include sunflowers, oats, rice, radish, alfalfa, cucumbers, and more.

Invasive garlic mustard secretes an allelochemical that disrupts the mycorrhizal network that supports many trees.

Now that we've examined some of the primary ways in which plants interact with each other, it's easy to see how this information can help us formulate strategies to positively affect the health of our cultivated plants. By considering how plants either cooperate with or inhibit one another, we can employ strategic plant partnerships to gain a desired result. Over the coming years, as we learn more and more about how the lives of plants, insects, and soil organisms intersect, we'll be able to develop additional techniques to influence the health of our plants in a positive way.

DIVERSITY = STABILITY

Though they are artificial environments, gardens are ecosystems where many layers of organisms interact with each other in myriad ways. As discussed in the beginning of this chapter, the perception of gardens as ecologically valuable habitats is pretty new. When built and maintained with an appreciation for all of the life it can support, a garden becomes so much more than a place to grow food and pretty flowers. It becomes an important environment, capable of filling some of the void left behind by the development and destruction of so many of our wild spaces. Gardens provide food, nesting sites, overwintering habitat, and other resources for many creatures when the gardener lets it be so. No other factor is as critical to a garden's ability to serve these functions as plant diversity.

Diversity refers to the number of different plant species present in a garden, while *structural complexity* refers to the assortment of growth habits and structures of those plants. When gardens are designed and planted with these two factors in mind, stability is the result.

Growing a diversity of garden crops and varieties in a mixed-planting environment creates a far more stable garden ecosystem than planting large swaths of a single plant species. We've known for quite some time that monocultures result in increased pest pressure, soil nutrient depletion, a lack of pollinators and pest-eating beneficial insects, and increased disease prevalence. In large-scale agriculture, monocultures make planting and harvesting easy, and growing large amounts of one variety means there will be consistency in the processed product made from that crop (tomatoes for sauce, cucumbers for pickles, or potatoes for fries, as a few examples). In contrast, vegetative diversity and structural complexity in a landscape create a more favorable environment for beneficial insects and biodiversity as a whole, at the same time creating a less favorable environment for pests and diseases. Diversity also prevents soil nutrient depletion and increases the variety of pollinators found in the area — not to mention the fact that growing a mixture of many different plants offsets losses when one species succumbs to a pest invasion or disease.

So how can companion planting help you have a more diverse, and therefore a more stable, garden with fewer pest and disease outbreaks? Well, in many ways, companion planting *is* planting for diversity. Although increasing your garden's diversity and complexity are probably not the original purpose of many of the companion planting techniques you might employ, they are definitely an added benefit. You may be pairing plants to provide a specific benefit to one or both of them, but an increase in plant diversity is the natural result.

On the other hand, some types of companion planting do have the specific goal of adding diversity to the garden. As you'll come to learn in chapter 5, companion planting with the aim of increasing plant diversity makes it harder for certain pests to discover host plants. It can also serve to repel or "trick" pests, as well as to disrupt their egg-laying behaviors. Whether intentional or not, improving your garden's biodiversity makes it a more stable — and better balanced — environment.

Southern green stink bugs mate on a stalk of pearl millet. Companion planting can make it more difficult for pests to find certain plants.

▸ **Companion planting is planting for diversity.**

SOIL PREPARATION & CONDITIONING

From Cover Crops to Living Rototillers

For generations, vegetable gardeners have headed out each spring, rototiller in tow, and tilled their garden soil into a fine, powdery texture prior to planting. Those who don't have a tiller turn their garden's soil by hand, with a spade or a border fork. This annual tradition is performed presumably to make the soil looser and more welcoming to plant growth. However, as we become more aware of the diversity of organisms within the soil and the complex ways in which these organisms help our plants grow, it becomes increasingly apparent that this kind of soil disturbance isn't necessarily in the best interest of our plants.

SOIL HEALTH *and* THE VEGETABLE GARDEN

Tilling or otherwise disturbing the soil disrupts the fungal colonies whose threadlike hyphae colonize the roots of plants and help them acquire nutrients and water in exchange for carbohydrates made by the plant. The workings of these mycorrhizal fungi were introduced in chapter 1. These beneficial organisms don't just play a role in the forest or prairie, they also influence the growth of our cultivated plants, whether they're in our foundation beds, woodland gardens, perennial borders, or, yes, our vegetable gardens.

Mycorrhizal relationships occur in two different ways. When the fungal hyphae penetrate the root cells themselves, they're called endomycorrhizae; when they're found primarily in the surrounding soil, they're known as ectomycorrhizae. With most woody plants, the mycorrhizal relationships are ectomycorrhizal, and you'll often find visible white, threadlike fungal hyphae spread throughout your mulch, topsoil, or pile of decaying leaves. The fungal networks found primarily in the vegetable and flower garden, however, are endomycorrhizal. These fungal partners enter the plant cells themselves, then extend out into the soil. Though you can occasionally spy fungal associations on the roots of your veggie plants, they are more often invisible to the human eye.

Just because we can't see the network of endomycorrhizae helping our vegetable plants acquire nutrients, it doesn't mean we can't nurture these relationships to the benefit of our garden. Healthy soils naturally contain plenty of fungal spores that will go on to germinate and infiltrate the soil and the roots of our plants. As discussed in chapter 1,

Healthy soil harbors billions of living organisms, the vast majority of which are invisible to the human eye.

mycorrhizae can connect the roots of different plants together and transfer nutrients between them, too. For annual plants, these networks may not at first seem as critical as they are for long-term woody plant communities, but that isn't the case. Not only do they transfer nutrients, these fungal hyphae help build good soil structure by "gluing" soil particles together or forming bridges between them, and by breaking down organic matter and releasing nutrients. Mycorrhizae may also help limit the occurrence of certain soilborne pathogens.

Regular additions of organic matter to the surface of garden beds improve soil without disturbing its structure or the established underground fungal network.

Running a rototiller over a garden bed pulverizes this delicate web of life within the soil, harms many different species of beneficial soil-dwelling insects, and, over time, destroys soil structure. Soil compaction negatively impacts the fungal network, too. While it's impossible not to disturb the soil when planting a seedling or a row of seeds, you should do your best to tread lightly and limit your digging to each individual planting site. Instead of tilling your vegetable garden each season, add an annual 1 to 2 inches of mulch to the top of it and leave the soil intact. Chopped leaves, leaf mold, or compost are among the best mulches for vegetable gardens.

Instead of tilling and digging, there are many natural ways to prepare and condition your soil to make it healthier and better able to support plant growth. Using the companion planting techniques presented in this chapter, you'll be able to improve soil structure and fertility, provide nitrogen, and even break up heavy clay soils.

There are three primary methods of utilizing companion plants for soil preparation and conditioning. The first uses cover crops to build healthy soil rich in organic matter; the second involves using leguminous plant partnerships for nitrogen transfer; and the final way utilizes plants with thick roots or root exudates to break up heavy soils. We'll discuss each of these concepts along with some specific plant partnerships you can employ to try them out. But before we dive into the first of these three methods, let me give you a general introduction to cover crops and, more specifically, how they can lead to better, healthier soil.

COVER CROPS *as* COMPANION PLANTS

Cover crops are nonharvested crops planted either before or after the harvest of a vegetable crop, or in fallow fields and gardens. Though they aren't planted and grown at the same time as the edible crop, utilizing cover crops is indeed a form of companion planting, since they can provide many benefits to your other plants. Previously perceived as useful primarily to farmers and those with very large gardens, the benefits of cover crops can be achieved even on a small scale.

Cover crops have many benefits when used appropriately, including:

+ Reduction in soil erosion
+ Creation of habitat for various species of pest-eating beneficial insects
+ Possible pest deterrence
+ Drawing up of nutrients from low in the soil profile closer to the soil's surface
+ Addition of organic matter to the soil
+ Nitrogen fixation, if the cover crop is leguminous (more on this in the next section)
+ Increased soil fertility
+ Improved soil structure
+ Increased biodiversity within the garden
+ Suppression of weed growth

I'll discuss many of these benefits here and there throughout this book, but for this chapter, I focus on using these plants to help improve the soil's structure and condition. In other chapters, cover crops are featured for their ability to support pollinators and other beneficial insects and for their helpfulness in managing weeds. As you'll come to see throughout the book, utilizing cover crops in your vegetable garden is a fruitful endeavor, no matter its size.

Typically planted in autumn or spring, cover crops are left to grow and then mowed down or turned into the soil before the desired crop is planted. When it comes to using cover crops to build healthy, productive soil, it's important to pair the correct species of cover crop with the right vegetable crop, the right season, and the right cover crop management technique.

Warm Season vs. Cool Season

Cover crops are typically divided into two categories: warm-season cover crops and cool-season cover crops.

Warm-season varieties are planted in spring or summer, either before a vegetable crop or in place of one in a fallow area.

Cool-season varieties are planted in late summer or early fall, after your veggies have been harvested. They must be planted early enough to germinate and grow before winter arrives. Some cool-season cover crops survive winter and regrow in spring, while others are killed by freezing temperatures.

PLANT PARTNERS *for* SOIL CONDITIONING

Dozens of different cover crops are available to farmers, but not all of them are appropriate for home gardens. Some can become invasive if they can't be turned under deeply enough. Others may reseed prolifically and become weedy in the garden if they aren't mowed at the exact appropriate time. That leaves a handful of cover crops that make excellent companion plants to improve the health of the soil in the home vegetable garden.

I've chosen the cover crops listed here for their ease of growth and availability. Not only do they add all-important organic matter to the soil, leading to improved moisture retention, they also bind soil particles together into rich, crumbly aggregates. In addition, decomposing cover crops feed the beneficial microbes that break down their organic matter into food for your plants. You can mix cover crops together, too, to reap the benefits of both a nitrogen-fixing legume and a biomass-providing cover crop that adds a lot of organic matter to the soil. For example, combine crimson clover with winter rye for a winter cover crop, or partner cowpeas and buckwheat for a warm-season cover crop.

Oats

This cool-season cover crop is winter-killed in climates with regular freezing temperatures, making it a great choice for those who like to plant early-season, cold-tolerant crops in the garden, such as lettuce, radish, peas, and brassicas like broccoli and cabbage. Oats (*Avena sativa*) may just be the best choice for those new to cover cropping because they are the most likely to die over the winter, and the following spring you can plant right through their residue. Yes, this helps limit weed growth (more on this in chapter 3), but most important, as the debris decomposes, it adds a hearty dose of organic matter to the soil. If you live in a warmer climate where oats do not die in winter, simply cut them down in spring when the plants come into flower. Oats can also be used as a living mulch (see page 59).

Oats are a great cover crop for beginners.

Buckwheat

Annual buckwheat (*Fagopyron esculentum*) is a warm-season cover crop that's best sown in spring, six to eight weeks before planting a warm-season vegetable crop such as zucchini, tomatoes, eggplants, or peppers. You can even plant it in between rows of corn; just be sure the rows are wide enough to get a mower between them to cut down the buckwheat when the time is right. Buckwheat can also be used as a soil cover in fallow beds during summer. Quick to germinate and grow, buckwheat should always be cut down just as it comes into flower since it's an aggressive self-sower if allowed to drop seed. Be sure to mow it within a week of when it starts to flower using a lawnmower or a string trimmer. Gardeners with small veggie plots can cut it by hand using a long-bladed lopper. Buckwheat trimmings can be left in place to serve as a mulch or turned into the soil to boost organic matter levels and feed beneficial microbes further down the soil profile. (See How to Use a Cover Crop, page 35.)

Winter Rye

A cool-season cover crop not typically killed by cold temperatures, winter rye (*Secale cereale*) is sown in fall and will resume growth in spring. It must be mowed just as it comes into flower to prevent the plant from reseeding. Don't wait for it to set seed or it will become weedy. Winter rye residue can be left in place as a mulch or tilled under. In chapter 3, you'll learn that winter rye residue is allelopathic, meaning it contains compounds that inhibit the growth of other plants, making it a good weed suppressor. But in terms of its soil-boosting power, the organic matter and burst of fast-decaying microbial food found in winter rye residues make it tops for helping build good soil fast. Quick growing and cold tolerant, winter rye is best for garden beds that will be planted with summer crops grown from transplants, like eggplants, squash, peppers, and tomatoes, since the residue may suppress the germination of vegetables grown from smaller seeds, especially if they're planted close to the time the rye is cut.

Buckwheat

Cold-tolerant winter rye seedlings poke through the snow.

Crimson Clover

Crimson clover (*Trifolium incarnatum*) is a cold-tolerant annual that can be used as a cool-season or warm-season cover crop. With a moderate growth rate, crimson clover is a legume, which means it's capable of taking nitrogen from the air and converting it to a form that can be used by other plants. A Pennsylvania study showed clover fixed enough nitrogen to supply the equivalent of 70 pounds of nitrogen per acre to a later crop. Not only is crimson clover exceptional at improving soil health and boosting soil nitrogen levels, studies have shown it to be an excellent nectar source for pollinators and other beneficial insects if it's left in the garden long enough to flower, something I'll discuss in chapter 8.

In areas with winters that don't dip much below −10°F (−23°C), crimson clover can be planted late summer through fall and cut down in spring, as soon as the plants bloom. In colder growing zones, crimson clover will not survive winter, and the residue can be left in place as a mulch. Crimson clover is not likely to regrow when mowing is timed properly. Consider using it as a winter cover crop and mowing it down prior to planting nitrogen-hungry crops such as corn and green, leafy vegetables. For use as a warm-season cover crop in fallow gardens, sow crimson clover seeds anytime in spring or summer, but be sure not to allow it to drop seeds. Crimson clover can also be used as a living mulch (see page 54).

Crimson clover

Young green sprouts of winter wheat poke through the soil.

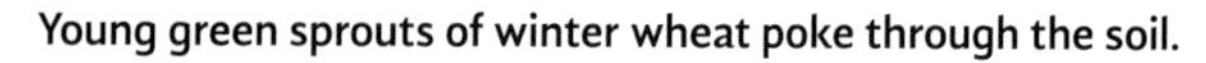

Cowpeas should be mowed just as they begin to flower.

Winter Wheat

Winter wheat (*Triticum aestivum*) is used as a cool-season cover crop. Planted in late summer through early fall, its rapid growth rate means it forms a protective winter soil cover, slowing winter soil erosion and helping to insulate the soil from dramatic freeze-thaw cycles. Winter wheat overwinters in warmer climates but is easier to kill via mowing the following spring than many other grain cover crops. Due to its extensive root system, winter wheat can be difficult to till under in spring, but when mowed at the time of bloom, there's no need to till, since the residue can be left intact; the plants are not likely to resprout. Vine crops such as pumpkins, winter squash, and cucumbers grow particularly well in the residues of mown winter wheat.

Cowpeas/Southern Peas

A warm-season cover crop best sown from early spring through summer, cowpeas (*Vigna unguiculata*) are excellent nitrogen fixers, providing up to 300 pounds of nitrogen per acre. They are fast to germinate, quick to grow, and drought resistant, providing excellent summer soil cover where necessary. Mow cowpeas when plants are a few weeks old and begin to produce flowers, before they drop seed. Cowpeas are not frost tolerant. Try growing cowpeas prior to planting a fall crop of leafy greens, such as lettuce, spinach, or kale.

When growing buckwheat as a cover crop, it's important to cut it down before the seeds are developed.

A gardener cuts down and chops up a green manure cover crop.

HOW TO USE A COVER CROP

Distribute cool-season cover crop seed over the garden soil toward the end of the growing season, after the harvest of desired crops. Warm-season cover crop seeds should be sown in early spring. There's no need to cover the seeds of cover crops with soil when planting. Allow the cover crop to grow until it reaches the flowering stage. Some cool-season cover crops may die over the winter, while others will continue to grow in spring. Mow or cut down the cover crop just as it comes into flower, and leave the trimmings in the garden. If you mow too early in the plant's development, it will regrow and try to bloom, so wait until the cover crop begins to flower to mow or cut it down as close to the ground as possible.

After mowing your cover crop, it's time to consider whether or not you should till the detritus into the soil. It's a decision I call the "tilling trade-off." As mentioned at the start of this chapter, tilling is a destructive process in many ways, but for some cover crops, tilling them under is the best way to ensure they're killed and won't resprout. Tilling also incorporates the cover crop residue into the soil where it can decompose, release nutrients, and provide food for soil microbes further down in the soil profile. Each garden (and gardener) is different, so to till or not to till cover crops is the choice of the gardener. I will say, however, that for winter-hardy cover crops that are not killed by a properly timed mowing, such as crimson clover, tilling may be your best option. For cover crops that are killed by cold temperatures or an aggressive, well-timed mowing, there's really no need to turn the debris into the soil.

With all cover crops, whether or not you choose to till them in, wait between two and four weeks to plant vegetables after the crops have been mowed or turned in. This mow-wait-plant cycle is very important, because it allows some of the cover crop residue to break down before the vegetable crop is planted. Cover crop residue in a fresh state stimulates a flurry of microbial activity that can be detrimental to plant growth. After a few weeks, this flush of activity will slow.

PLANT PARTNERS *for* NITROGEN TRANSFER

The second way companion plants can be used to condition the soil is by using living plants as nitrogen generators. Nitrogen is one of three macronutrients needed by all plants. It's an important component of proteins and of the chlorophyll molecule, for example. Nitrogen supports green, leafy growth. Vegetable crops that produce edible leaves or grow very tall (such as lettuce, spinach, cabbage, kale, and corn) make use of more nitrogen than crops that produce fruits, such as cucumbers, tomatoes, peppers, and watermelon. That's not to say that fruiting crops don't need nitrogen; they certainly do, but leafy crops tend to require more nitrogen to maximize yields. Nitrogen also happens to be very volatile, meaning it dissipates rapidly. It moves quickly within the soil, and common forms of nitrogen, such as nitrate and ammonia, leach easily from the soil, making it all too easy for soils to become deficient in this important nutrient.

While nitrogen is the most plentiful element in Earth's atmosphere, atmospheric nitrogen is always found as two N molecules stuck together (N_2). Most plants can't use atmospheric nitrogen to fuel their growth, so instead they have to get it from the soil. Fierce competition for nitrogen between plants and soil microbes can further exacerbate deficiencies. If the lower leaves of a plant are turning yellow, it may be due to nitrogen deficiency because nitrogen-hungry developing leaves will draw it away from older leaves.

Members of the pea and bean family (Fabaceae), along with a handful of other plants, have the unique ability to take nitrogen from the air and transform it into a form plants can use. It's a process called nitrogen fixation. For this process to occur, leguminous plants require a partnership with a specialized bacteria (primarily rhizobia). The bacteria colonize the roots of these plants and form nodules on them. Inside these nodules, in exchange for sugars produced by the plant, the bacteria fix nitrogen into compounds that their host plant and other nearby plants can use to grow, develop, and reproduce.

When a leguminous plant dies and decomposes, the usable nitrogen within that plant and the nodules on its roots becomes available to future plants grown in the same area. This is why farmers and gardeners often use members of the pea and bean family as green manures. But nitrogen-fixing plants can share nitrogen while the plants are still in a living state. This nitrogen is shared in a few different ways. It can enter the soil via root exudates or by the frequent death of nodule-containing roots throughout the growing season. By now you shouldn't be surprised to learn that nitrogen fixed by a legume can be shared with neighboring plants via the mycorrhizal fungal network, which accounts for 20 to 50 percent of nitrogen transfer from legumes to nonlegumes. One study found that, in total, leguminous plants can transfer between 30 and 50 pounds of nitrogen per acre.

These types of living plant partnerships can be used to improve plant growth and yields, increase soil fertility, and reduce the use of commercial fertilizers. Interplanting crops that cannot fix their own nitrogen with leguminous plants capable of fixing and sharing nitrogen can be a real win for gardeners. The process of nitrogen fixation may be invisible to the gardener's eye, but it impacts plant health and garden production in a big way.

Some leguminous garden crops are better at fixing nitrogen than others. While traditional garden beans and peas fix significantly less nitrogen than legumes like cowpeas, fava beans, and soybeans, they still make useful companion plants when partnered with other vegetables that are low to moderate nitrogen feeders. The following companion planting strategies are aimed at reducing the need for supplemental nitrogen fertilizer and naturally generating high yields of healthy veggies.

For any of the following plant partnerships to be effective, the two plants must be in close proximity to each other, and the soil should remain undisturbed during the growing season. At the end of the gardening season, after both crops have been harvested, allow the leguminous plants' residue to break down in place to deposit any remaining nutrients into the soil.

The symbiotic relationship between leguminous plants and the bacteria that colonize them happens only in these specialized root nodules. When crushed or cut open, the nodules are red inside due to the presence of leghemoglobin, the protein that keeps oxygen levels balanced within the nodules (much like the hemoglobin in our own blood).

Garden Beans + Potatoes

A 2010 study found that partnering potatoes (*Solanum tuberosum*) with common garden beans (*Phaseolus vulgaris*) such as green beans and yellow wax beans produced an increased potato tuber size. While garden beans don't fix large amounts of nitrogen, they are capable of sharing some of the nitrogen they do fix with nearby plants via the mechanisms described above.

This partnership can be implemented by alternating rows of beans and potatoes in the garden or by mixing the two plants together in the same row or block. Regardless of the style you choose, the planting of potatoes and beans can occur simultaneously, or the potatoes can be planted several weeks in advance of the beans. "Seed" potatoes can be planted any time from early spring through midsummer in temperate climates. Bean seedlings take several weeks to begin the nitrogen fixation process, and newly planted seed potatoes take a few weeks to begin to grow, making this partnership both flexible and straightforward.

Fava Beans + Sweet Corn

Fava beans (*Vicia faba*) are among the best nitrogen-fixing legumes. They fix up to 250 pounds of nitrogen per acre. Fava beans have long been used as a nitrogen-fixing cover crop. In one study, farmers of sweet corn (*Zea mays* var. *saccharata*) halved their fertilizer needs by using a fava bean cover crop prior to planting the corn, and the researchers noted that other nitrogen-hungry crops could benefit from a fava bean partnership, too. But favas aren't limited to use as a cover crop. They can also transfer nitrogen to companion plants while they're still in a living state.

Like garden peas, favas are cool-season plants that prefer to be planted in very early spring rather than in summer. The plants are tolerant of frosts and grow best at a temperature 60° to 65°F (16° to 18°C). Since it takes legumes like fava beans time to develop their relationship with nitrogen-fixing bacteria and begin providing nitrogen to other nearby plants, the timing of this companion planting strategy works well. First, the beans are planted in early spring. Then, when the plants are a few weeks old and the threat of frost has passed, the corn is planted. The favas will continue to grow and share nitrogen with the corn until they're ready for harvest about 100 days after planting. After harvest, leave the plants in place. Their roots and shoot residues will add further nutrients to the soil as they decompose.

Alternate crop rows or plant two adjacent rows of fava beans, leaving a 10-inch-wide strip of soil between them where the corn will be planted. Plant several of these "sandwich rows" in the garden since corn is wind pollinated and many plants are needed for good pollination to occur.

Cowpeas + Peppers and Other Tall Transplants

Cowpeas (*Vigna unguiculata*) are an excellent warm-season companion plant for nitrogen fixation and can be used as both cover crops and living companion plants. Drought resistant, cowpeas have a long taproot that can obtain moisture from deep in the soil and draw it up closer to the surface.

A California study showed that cowpeas improved production in peppers by reducing weeds and providing nitrogen. Cowpeas are somewhat allelopathic, so they can negatively impact the germination of smaller seeds. Because of this, it's important to use cowpeas as a companion plant with crops that are planted as transplants, not grown from seed. Cowpeas are also useful as a living mulch for weed control (see page 60).

Cowpeas are best planted in spring when the threat of frost has passed. Transplants of larger-statured vegetable crops, such as tomatoes, peppers, summer squash, and eggplants, can then be planted alongside the cowpea seedlings. If you'd like to harvest your cowpeas, select a cultivar that's good for eating (black-eyed peas, for example) rather than one that's better suited for use as a cover crop.

Peas + Lettuce

This plant partnership works for many different reasons. First, both garden peas (*Pisum sativum*) and lettuce (*Lactuca sativa*) are cool-season crops that grow best in spring or fall. Second, they have different statures that partner well together: peas produce tendrils and climb, while lettuce stays low to the ground. When early summer arrives and the weather heats up, lettuce is triggered to go to flower (a process called bolting), but the shade provided by the pea vines can prolong the harvest by delaying the bolting process.

Peas also benefit lettuce by providing them with nitrogen via the mycorrhizal network, root exudates, and more. As mentioned above, lettuce is a green that requires adequate nitrogen levels to produce thick, succulent leaves. To employ this partnership, flank each row of peas with a row of lettuce on either side, or alternate rows of these two crops, using a trellis or fence to support the growing peas.

Edamame + Fall Greens

Soybeans (*Glycine max*) are an often-used cover crop on large farms, grown as both a food and commodity crop. But did you know that those fuzzy green pods you enjoy at your favorite sushi restaurant are actually immature soybeans? Yep, that's right. Edamame are immature soybeans, though the soybean varieties grown for edamame production are selected for their flavor, while those grown for dried bean and/or cover crop production are selected for other traits. All varieties, however, fix nitrogen.

Edamame are soybeans harvested in a young, green state, and since they're a member of the legume family, they are among the best nitrogen fixers. When used as a cover crop, they provide about 130 pounds of nitrogen per acre.

Soybean plants begin to fix nitrogen when they are a few weeks old and continue to do so for a few months thereafter. This makes edamame planted in spring, after the danger of frost has passed, a good companion plant for a late-summer sowing of cool-season greens such as kale, collards, spinach, and chard. Alternate crop rows to help ensure that the nitrogen produced by the soybeans is available to nearby greens.

Inoculating Legume Seeds Prior to Planting

The rhizobia bacteria that colonize the roots of leguminous plants and enable them to fix nitrogen are present in most healthy garden soils, especially where legumes have grown before. But purposely inoculating leguminous crops with a powdered or liquid form of these bacteria at the time of planting can speed up the rate of root colonization as well as increase the amount of nitrogen fixation that occurs. Pea and bean inoculant, a microbial amendment composed of millions of live bacteria, is readily found at garden centers and in seed catalogs.

Different inoculants are available. For garden peas and beans, the best species is *Rhizobium leguminoserum*. Other species of microbial inoculants, such as *Sinorhizobium meliloti* and *Bradyrhizobium japonicum*, are useful for alfalfa, clovers, soybeans, and other members of the legume family.

There are several ways to introduce legume inoculant products to your seeds prior to planting, but the best method for applying each different type of inoculant will be specified on the packaging.

PLANT PARTNERS *for* BREAKING UP HEAVY SOILS

The third and final way companion planting can be used to prepare and condition your soil is by cultivating plants with thick roots or root exudates to help break up heavy soils. When used as cover crops, the companion plants featured here will improve soil structure by adding organic matter and nutrients as they decay, but in this case, we grow them with the specific goal of breaking up heavy soils; the nutrients and organic matter they add are side benefits of the process. Here are three plant partners that can be used in this manner.

Buckwheat

Though I've already mentioned the power of buckwheat to add organic matter and nutrients to the soil, it's also capable of loosening heavily compacted soils via the root exudates it produces.

Plant root exudates consist of a complex concoction of sugars, amino acids, enzymes, vitamins, and many other compounds. These exudates have both indirect and direct effects on how plants acquire nutrients to fuel their growth. The area around the root, known as the rhizosphere, is filled with microorganisms that interact with these root exudates in many ways. The bacteria that form nitrogen-fixing nodules on plants react to the presence of these exudates, the development of mycorrhizal fungi is influenced by root exudates, and plants can even use root exudates to foster certain soil microbes that help them procure nutrients and to change the soil's pH in their rhizosphere. These exudates can alter the chemical and physical properties of the soil, too.

When buckwheat is used as a cover crop in areas with heavy soils, its root exudates alter the way soil particles stick together and open up channels within the soil for water and nutrients to move around in. In addition, the microbial communities found in the rhizosphere of buckwheat plants can further improve the structure of certain types of soil. Buckwheat produces allelopathic compounds, so it may limit weed growth as well, something I'll address further in chapter 3.

Buckwheat roots are great at loosening heavily compacted soils, but the plants must be cut down before they go to seed.

The long, thick taproot of the forage radish serves as a living rototiller.

Forage Radish

These are not the same radishes you'll find in your salad mix. These long, tapered radishes are instead prized for their prowess at biodrilling, a process that uses cover crops to break up compacted soil to improve water filtration and gas exchange. Because of their long, thick roots, forage radishes (*Raphanus sativus* var. *longipinnatus*) alleviate compaction and serve as an excellent "living rototiller" of sorts.

Taprooted plants like forage radish exert pressure and push through compacted soils to open up channels and improve the soil's friability (the ease with which it crumbles and fragments from larger clods into smaller ones). The roots of forage radish can grow several feet deep and create large vertical openings in the soil. They also have long, fine root hairs that extend as deep as 6 feet into the soil from the base of the thicker portion.

An added benefit of forage radishes is that, as they grow, they pull up nutrients from deep within the earth. Then, after the plants are winter-killed and their residues begin to decompose in the garden, those nutrients are returned to the soil much closer to the surface, where subsequent crops have easy access to them. The roots also leave behind open soil channels. Forage radish are winter-killed in regions where the temperature regularly dips below 20°F (–7°C). In warmer climates, the plants won't die until after flowering the following spring, so forage radish isn't a good fit for those areas.

To use forage radish as a companion plant cover crop to break up heavy soils, it's important to select the right varieties. Forage radish varieties like Groundhog, 'Graza', 'Sodbuster', Tillage Radish, and 'Eco-Till' are good choices and less expensive to use than daikon-type radishes that were bred as an edible crop. Seeds are sown in late summer or early fall, about six weeks before regular frosts arrive, and the roots are left to grow until they're winter-killed. Radishes decompose quickly, so the area will be ready for planting come spring.

This is a particularly good companion planting practice for no-till gardeners because it serves to reduce compaction without destroying the soil biology and mycorrhizal fungal network. Do not sow the seeds too thickly, or smaller roots will be the result. For home gardens, plan on using a seeding rate equivalent to about 6 to 8 pounds of seed per acre. Studies have found that biodrilling improves root growth in vegetables later grown in the same area. The veggies are also more resilient to drought conditions.

Turnips

While you may not be a big fan of turnips (*Brassica rapa* subsp. *rapa*) on your dinner plate, they're very useful for alleviating soil compaction. Turnips are an excellent cool-season cover crop for breaking up heavy soils. Most turnip varieties do not have long, deep taproots like the forage radish, but special varieties have been bred to produce an elongated root that's below ground level, unlike the turnips we eat, which have a bulbous root that protrudes above the soil surface. 'Appin' is a long, tapered turnip that's perfectly suited to companion planting for biodrilling.

To use turnips as a biodrilling cover crop, plant in late summer, about six weeks before hard freezes arrive. Sow at a rate of 1 to 2 pounds of seed per acre for home gardens. Turnips survive winter temperatures down to about 20°F (–7°C), making them not much hardier than forage radish. If your turnips survive the winter, kill them by mowing the plants down just as they come into flower. The residue will take a few weeks to break down, then the area can be planted with a harvestable warm-season crop.

Turnip varieties that produce long, tapered roots are great for opening up compacted soils.

WEED MANAGEMENT

Using Living Mulches and Allelopathy to Combat Weeds

Companion planting can reduce weeds in the vegetable garden in two primary ways. The first is by employing living mulches to help crowd or shade out competing weeds. The second is through allelopathy, which we touched on in chapter 1. The growth-inhibiting chemical compounds produced by certain plants can be a huge help to gardeners combating weeds. We'll look at the best living mulch companion plants as well as some specific allelopathic plant partners that are useful for controlling weeds around certain crops.

Many living mulches, including this red clover, support pollinators and other beneficial insects. When they're in flower, they provide nectar and pollen. When they're not in flower, they provide excellent habitat.

LIVING MULCHES

In essence, living mulches are an extension of cover cropping, except living mulches are grown around or between rows of actively growing, harvestable crops, rather than in fallow garden spaces. Living mulches have been shown to reduce weed growth significantly while they're in a state of active growth, though the extent of the control is highly variable and depends on many factors. In addition to competing with weeds for available resources, living mulches block light from reaching the seeds of potential weeds, preventing them from germinating. Like cover crops, living mulches reduce soil erosion, increase beneficial insect habitat, and enhance biodiversity. If a flowering living mulch plant is used, it may draw pollinators. Living mulches have been proven to foster a diversity of beneficial predatory and parasitoid insects that help control pests. Finally, as an added bonus, living mulch plants that are leguminous supply nitrogen to their partners, as discussed in the previous chapter.

There is great research that points to yet another benefit of interplanting vegetables with living mulches: this kind of mixed-planting matrix increases the mycorrhizal population in the soil and leads to improved yields and plant health. The mycorrhizal network is more extensive in mixed plantings than in monocultures, giving partner crops improved access to nutrients and water. Not to mention that multiple studies have shown a garden with a more diverse selection of plants exhibits a reduced vulnerability to pest outbreaks. We'll take a deeper look at how companion planting affects pest populations in chapter 5.

Typically, low-growing plants make the best living mulches because they're less likely to compete with the taller crops they're partnered with.

How to Use a Living Mulch

Living mulch companion plants are used in a handful of ways:

+ Rows of living mulch plants alternate with vegetable crop rows throughout the garden.
+ Living mulch plants are grown directly underneath the skirts of taller crops in planting rows or blocks.
+ Living mulch plants serve as walking paths throughout the garden, instead of traditional mulches like straw or shredded bark.
+ Narrow strips of living mulch plants flank each block or row of vegetable crops.
+ Living mulches are sown throughout orchards beneath and around the trees.
+ Living mulches are planted around the base of grapevines and between rows in vineyards.

Finding the perfect living mulch plant partner for a particular crop does require forethought. For example, a perennial living mulch such as common thyme (*Thymus vulgaris*), white clover (*Trifolium repens*), or alfalfa (*Medicago sativa*) makes a better companion for permanent fruit tree or vineyard plantings than for an annual vegetable crop. In general, annual crops should be partnered with annual living mulches and perennial with perennial, though this isn't a hard-and-fast rule. There are a handful of perennial living mulch plants that work well, especially when used between planting rows or in garden walkways, but it's important to use them correctly so they don't spread beyond bounds or compete with their companions.

A living mulch of crimson clover is a great addition to a home vineyard. Not only does it deter weeds, it also lures in pest-eating beneficial insects.

The mature height of the living mulch is another factor to consider. Tall vegetable crops such as tomatoes and peppers, vertically grown vine crops, and fruit and nut trees won't be outcompeted by slightly taller living mulches, unlike a small-statured vegetable such as lettuce or radish. Finally, coupling a living mulch with a vegetable crop that requires similar growing conditions is a must, though there is some flexibility when it comes to partnering a cool-season living mulch with a warm-season vegetable crop or vice versa.

To grow living mulches, plant seeds according to their preferred growing season using whichever interplanting method is best for your particular garden area. When using living mulches in the vegetable garden around and between harvestable crops, the seeds should be sown either slightly before or slightly after crop transplants go out into the garden to limit competition. When using living mulches under perennial crops, like berries, grapevines, and fruit trees, you should sow annual living mulch seeds in spring and perennial living mulch seeds in either spring or fall.

For the best management, occasionally mow or otherwise cut back living mulches, especially leguminous ones, to return the nutrients in their shoots to the soil via decomposition and to keep them from competing with your vegetable crop. The first mowing should take place soon after the living mulch reaches the canopy height of the transplanted vegetable seedlings, giving the vegetable transplants time to outgrow them. For example, if you grow crimson clover as a living mulch around tomato plants, mow or cut back the clover a few times throughout the season, starting from the time the clover and the tomato transplants are the same height, and leave the trimmings in place to break down and provide nitrogen and other nutrients to the growing tomato plants.

New buckwheat seedlings sprout from the soil.

Potential Stumbling Blocks

There are, of course, a few drawbacks to using a living mulch. Though living mulches are meant to outcompete weeds, they can generate too much competition with their vegetable crop partners for light, water, and nutrients, making a thoughtful partnership absolutely essential. Also, some living mulch crops throw a lot of seed, causing them to become a weedy nuisance if the plants aren't mowed soon after they come into flower. Regular mowing is essential to manage living mulches, regardless of whether they're grown between crop rows or underneath crops. Using a lawnmower, a string trimmer, or a manual tool like a scythe, sickle, or hand loppers to complete this task requires time and energy.

If you are worried about competition between the living mulch plant and its companion crop, don't interplant the two in the same area. Instead, opt for using the living mulch in walkways or in dedicated rows placed between vegetable crop rows. These methods limit competition, though the living mulch will still need to be appropriately mowed or trimmed when it grows too high.

PLANT PARTNERS *as* LIVING MULCHES

Despite the thought and maintenance required when using living mulches, they're a huge boon to the health of your garden. Let's look at specific companion planting combinations that utilize living mulches for the betterment of the partner crop. Some of these choices not only suppress weed growth through physical competition, but they also produce allelopathic compounds, offering a multipronged approach to weed management.

Crimson Clover + Cole Crops

Crimson clover (*Trifolium incarnatum*) is a cold-tolerant annual legume often used as living mulch as well as a cover crop. It can grow quite tall, but regular mowings keep it tamed. Crimson clover suppresses weeds by forming a thick mat, and it supports high densities of beneficial insects by providing food and habitat. Numerous studies extol the benefits of using crimson clover around cole crops such as broccoli, cabbage, Brussels sprouts, kale, and cauliflower. Use it either between crop rows or as a ground cover around the plants. Be sure to mow crimson clover before it has a chance to drop seed. Where winters consistently dip below 0°F (–18°C), crimson clover is winter-killed.

Medium Red Clover + Winter Squash

Medium red clover (*Trifolium pratense*) helps suppress weeds and breaks up the soil. It's a short-lived perennial that lives for about two years in the South. It's more of a winter annual in northern zones. Like other legumes, medium red clover fixes nitrogen and provides it to nearby plants in both a living state and when the plants are left to decompose in the soil. The extensive root system of medium red clover holds the soil in place, and the thick plants provide habitat for beneficial insects. It can be cut multiple times and grows back quickly, making it a great choice for a living mulch.

Medium red clover partners well with winter squash. Because it's a vigorous grower, medium red clover outgrows weeds in winter squash patches while still allowing the vines to ramble over it. In studies, the best results were achieved by alternating strips of clover with rows of winter squash.

White Clover + Strawberries or Blueberries

Combining strawberry plants (*Fragaria × ananassa*) with a living mulch of white clover (*Trifoleum repens*) provides many benefits. In this case, the living mulch is planted between strawberry rows rather than interplanted with the strawberries themselves. Several living mulches have been trialed with strawberries in research. The living mulch strips between rows were mowed just before strawberry harvest. White clover was found to be one of the best covers because it continued to grow after mowing. And, after mowing, it decomposed faster than the other living mulches trialed. Keep in mind that the clover and strawberries should not be planted together in a mixed planting, since that was shown to have negative effects due to increased competition.

Similar results were found when blueberry farmers used white clover as a living mulch around blueberry shrubs. It doesn't need to be mown as frequently as grass, and the nitrogen it fixes goes on to feed the blueberry plants.

White Clover + Tomatoes, Peppers, Eggplants, and Other Tall Vegetables

White clover is a great choice for a permanent living mulch between rows of vegetables, small fruits, and orchard trees. White clover is a low-growing legume that stays in place year-round due to its perennial growth habit. It is winter hardy down to about −30°F (−34°C). Several types of white clover can be used as a living mulch, each with a different mature size. 'Wild White' clover is very low growing and can be walked on repeatedly without ill effects, 'Dutch White' and 'New Zealand White' are moderately sized and flower earlier than some other varieties. Larger types of white clover reach up to 12 inches tall and don't typically make good living mulches because they may outcompete the vegetable crop.

One study found that when white clover was used as a living mulch between rows of various vegetable crops, the weed control it provided was comparable to commercial herbicide applications. It is more shade tolerant than most other living mulch choices, making it a good fit for planting near taller vegetable crops like tomatoes, peppers, and eggplants. White clover is easy to mow to keep it from setting seed. It's worth noting that white clover has been shown to reduce levels of cabbage aphids in broccoli in a California study, thanks to its ability to serve as a host plant for aphid predators and parasitoids.

Subterranean Clover + Many Vegetables

Subterranean clover (*Trifolium subterraneum*) is a winter annual legume. Like peanuts, fertilized flowers of subterranean clover form pegs that grow down and penetrate the soil; the seeds are then formed below the ground. When used as a living mulch beneath several different crops, including sweet corn (*Zea mays* var. *saccharata*), summer squash (*Cucurbita pepo*), cabbage (*Brassica oleracea* var. *capitata*), snap beans (*Phaseolus vulgaris*), and tomatoes (*Lycopersicon esculentum*), subterranean clover was found to provide excellent weed control, and the yields of these crops were not negatively affected by the mulch. Weed biomass was substantially lower than control beds without the clover.

Subterranean clover is not winter hardy in northern zones that drop below 15°F (–9°C). In California, it's a common year-round living mulch in almond orchards, where it self-seeds and persists through winter. Often called subclover, this legume is low growing and an exceptional nitrogen fixer. Its thick mat of stems and leaves outcompetes weeds when planted between vegetable crop rows. In one Maryland study, a subclover living mulch controlled weeds better than conventional herbicide treatments.

Like other living mulches, mow subclover throughout the growing season to prevent competition with desired crops and to keep the pegs from reaching the ground and developing seeds. Various studies have partnered subclover with broccoli, cauliflower, lettuce, sweet corn, cabbage, leeks, and other crops. Keep in mind, though, that subclover does have allelopathic effects, so do not use it when planting vegetables from seed.

Oats + Tall Vegetable Crops or Berries

Oats (*Avena sativa*) make an inexpensive, reliable living mulch that works well around many different vegetable crops, especially taller varieties and those trained to grow up a trellis. The quick growth and excellent weed-suppressing abilities of oats, coupled with the fact that oats are easily killed by cold winter temperatures, make this grain a prime living mulch for use around fall-planted cool-season crops, too. The biggest downside of using oats as a living mulch is the size of the plants. Oats can grow up to 4 feet tall, so regular mowings are a must. In southern regions where summer weather is hot and dry, oats may not hold up well.

Oats smother and outcompete weeds, plus the residue left behind after mowing is allelopathic, meaning it hinders future weed seed germination (more on this in the next section). Oats are best partnered with crops grown from transplants, not from seed.

Much of the baled straw gardeners use to mulch their garden is made from oat stems, so growing oats as a living mulch and regularly mowing it means you won't have to purchase bales of straw to spread between crop rows as weed control. Oats also make a great living mulch between strawberry rows in northern growing zones. When sown in late summer or early fall, the oats grow for several weeks before being killed by freezing temperatures. Come spring, the oat stems can be rolled or flattened down to the ground, forming a mat of dead stems over the soil between the strawberry rows. The technique works well in row plantings of bramble fruits, such as blackberries and raspberries.

Winter Rye + Asparagus

Living mulches growing below the fern canopy of an asparagus bed provide several benefits, including weed suppression and soil protection. Researchers found that planting winter rye (*Secale cereale*) after the asparagus harvest season ends in late spring reduced weed pressure significantly, particularly for fall-germinating weeds like dandelion. Winter rye that's planted in late spring grows quickly but not as tall as plants grown from seed sown in fall. This is because the exposure to winter temperatures in fall-planted rye causes the rye to grow to a taller height. The late-spring planting of rye in an asparagus patch naturally dies back with the arrival of summer's heat, keeping it from competing with the asparagus ferns. The caveat of this study, however, was that the researchers found that the asparagus plots interplanted with a living mulch of winter rye required more irrigation throughout the growing season. To prevent problems, be sure to keep the asparagus patch well watered.

Cowpeas + Peppers

Cowpeas (*Vigna unguiculata*) form a thick mat of roots and a canopy of leaves capable of restricting weed growth. They are mildly allelopathic and can impact the germination of smaller weed seeds. One particular study showed improved growth and production of a pepper crop when grown in conjunction with a summertime living mulch of cowpeas.

Yellow Mustard + Summer Squash

Yellow mustard (*Sinapis alba*) is an annual cover crop that when used as a living mulch has been shown to increase the yields of summer squash. The same study also noted that this plant pairing decreased the densities of certain pests and diseases, such as aphids, whiteflies, and squash silverleaf.

Mustard living mulches can inhibit small-seeded annual weeds from germinating, and they are very effective at suppressing winter weeds, too, if the winter-killed stems are left in place. Yellow mustard is sensitive to cold temperatures and is winter-killed at about 25°F (–4°C). Used as either a spring or summer living mulch, or as a fall-grown living mulch, it provides fast cover that shades out weed seeds and outcompetes them.

The thick, powerful root system of yellow mustard makes it a great partner for relieving soil compaction and increasing water infiltration. Like many other members of the brassica family, if allowed to produce flowers, it produces blooms that are very attractive to pollinators and other beneficial insects. Keep in mind, however, that yellow mustard can become highly invasive if the plants are left to drop seed, making it important to mow them down before the seeds develop.

PLANT PARTNERS *for* ALLELOPATHY

Another way plant partnerships reduce weeds is by the production of allelopathic chemicals. (A brief rundown of these compounds can be found on page 15.) The allelochemicals produced by certain plants may be exuded through their roots, leached from their leaves or shoots, or released through the process of decay as their residues break down in the soil. Allelochemicals can also influence or alter the microbial activity within the soil, impacting other plants in an altogether different way.

No matter how these compounds are released or the extent of the impact they have on soil biology, when an allelopathic plant is used as a cover crop or living mulch, it has the potential to negatively influence the growth of other plants. That's what they've evolved to do, after all. Perhaps they reduce germination rates, inhibit seedling growth, or outright kill other plants. Certainly not all plants produce allelopathic chemicals, but among those that do, each specific species contains a different concoction of allelochemicals that affect other plants in different ways.

Many studies document the allelopathic effects of various plants. The effects of some allelopathic cover crops last for days, weeks, or months after the plant dies, while the effects of others are short lived, persisting in the soil only while the plant is alive. While we aim to make companion planting partnerships that suppress weeds, sometimes the desired vegetable can be suppressed, too. All of this means that using allelopathy to combat weeds is a complicated affair. If done with care and consideration, however, it's extremely effective.

Allelopathic interactions are often very specific. For example, allelopathic rye grass is excellent at suppressing certain annual weeds, including lamb's-quarter, purslane, and crabgrass, but not others. So, too, with allelopathic subclover and cowpeas.

Some plants are more sensitive to allelochemicals than others. Lettuce and cabbage transplants are extremely sensitive to certain allelochemicals; tomatoes, cucumbers, and eggplants are much less so. The size of the seeds seems to matter, too: when it comes to vegetable crops grown from seed, larger-seeded crops show a greater tolerance than crops with small seeds.

Some common cover crops are allelopathic. Leaving these cover crop residues on top of the soil as opposed to tilling them in impacts the extent of the allelopathy. Tilling causes the allelochemicals to break down more quickly, while leaving the residues as mulch extends their weed-suppressing abilities for much longer. Allelopathic cover crops previously discussed in this book include the following:

+ **Winter rye.** If you have a long-standing weed problem in your garden, winter rye may be your answer. Rye residue has been shown to reduce foxtail, pigweed, ragweed, and purslane germination between 43 and 100 percent, to name just a few of the weeds it helps manage. When sown as a cover crop in late summer or autumn and mowed or cut the following spring, it provides weed control for months if the residues are left in place as a mulch rather than tilled in. Essentially, any vegetable that's grown from transplant rather than from seed is a perfect plant partner for winter rye.
+ **Oats.** The allelopathic properties of oats help manage weed growth when oats are grown prior to a harvestable crop and the residues are left behind. In sweet potato cultivation, a cover crop of oats followed by a mulch of oat stems significantly reduces weeds.

The power of allelopathy and living mulches to manage weed growth is one of the core concepts of science-backed companion planting, and it's a highly useful one at that. With a little experimentation and research, finding perfect plant partners to reduce weed woes pays big dividends, both in terms of a reduced workload and herbicide inputs, not just on large farms but also in backyard gardens. Like the allelopathic cover crops already discussed, the plant partners described on the next two pages can be used to limit weed growth in your vegetable garden.

Bottle gourds grow on a mulch of oat residue.

Cucumbers + Taller Vegetables

You might be surprised to hear that cucumbers (*Cucumis sativus*) are allelopathic. They produce several growth-inhibiting allelochemicals that negatively impact neighboring plant species, cinnamic acid being among the most studied. In addition, cucumbers are autotoxic, which means their allelochemicals can negatively impact members of their own species if the plants are spaced too closely. Watermelons produce autotoxins, too, as do several other members of the cucumber family (Cucurbitaceae).

Cucumbers can be used as a weed-management tool in a couple of ways. They can be grown in weed-plagued gardens as a thick ground cover. Then their residue can be tilled into the soil prior to planting a late-season crop three to four weeks later. Cucumbers can also be grown as a living mulch under or around taller crops such as corn, tomatoes, okra, eggplants, and the like to help limit weed growth. The allelochemicals are exuded through their roots, and these allelochemicals have been shown to be even more toxic when the cucumber plants reach reproductive age. As with rye, it's best not to use cucumbers as a weed control where growing other crops from seed.

Rapeseed + Potatoes

If your potato patch is overrun with weeds, a cover crop of rapeseed, also called canola (*Brassica napus* var. *oleifera*), may do the trick. One study examined the allelopathic properties of rapeseed and determined it to be an effective weed control in potatoes (*Solanum tuberosum*). Like yellow mustard, rapeseed is a member of the brassica family and should not be allowed to set seed or it will become invasive.

Rapeseed's extensive, thick root system helps loosen topsoil as it brings nutrients closer to the soil surface. Annual varieties of rapeseed are killed by freezing temperatures. Biennial varieties that are planted in fall and flower the following spring are not killed by cold temperatures and must be mowed down when the plants come into flower, three to four weeks prior to planting the potatoes.

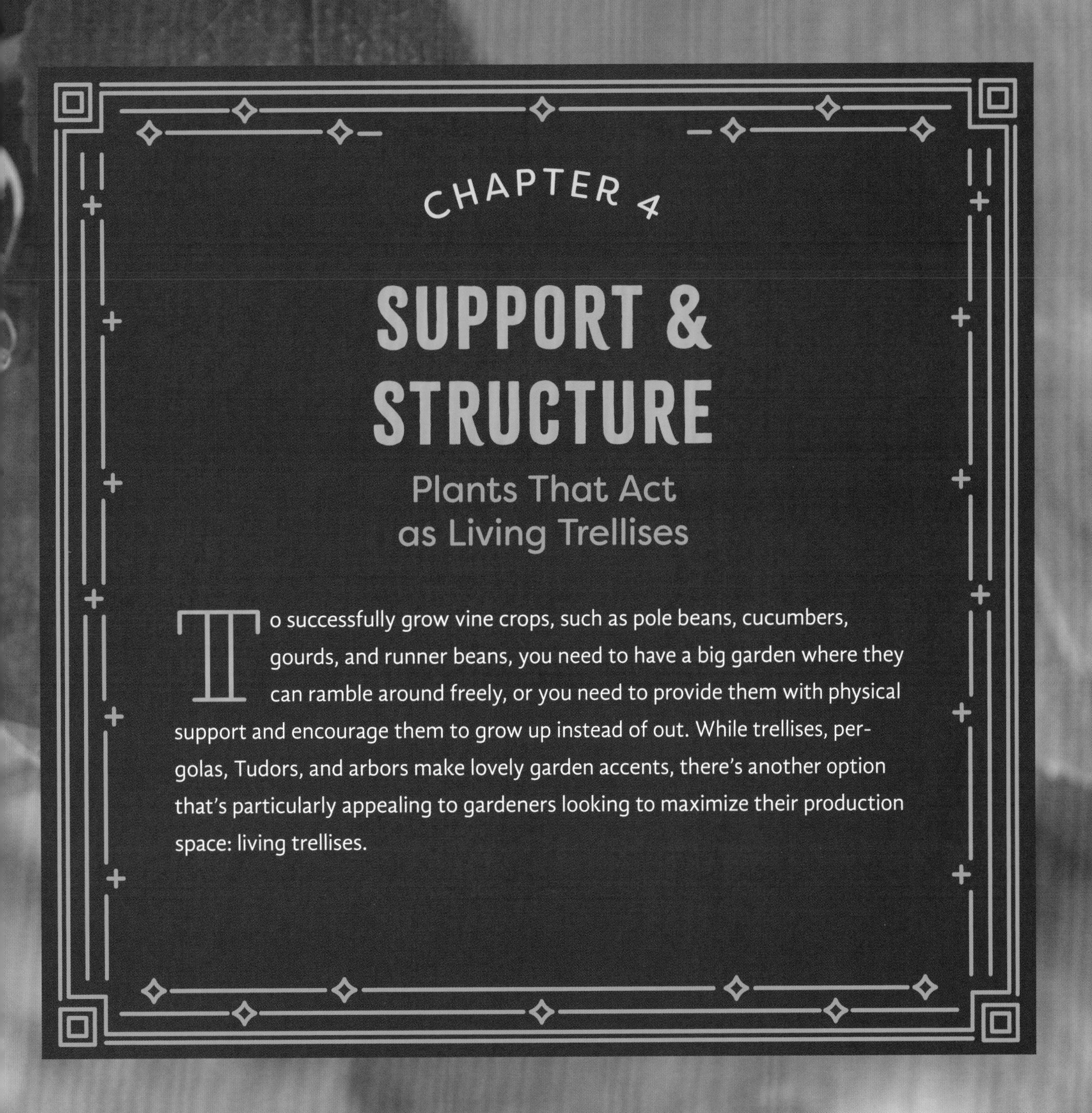

CHAPTER 4

SUPPORT & STRUCTURE

Plants That Act as Living Trellises

To successfully grow vine crops, such as pole beans, cucumbers, gourds, and runner beans, you need to have a big garden where they can ramble around freely, or you need to provide them with physical support and encourage them to grow up instead of out. While trellises, pergolas, Tudors, and arbors make lovely garden accents, there's another option that's particularly appealing to gardeners looking to maximize their production space: living trellises.

PRACTICAL BEAUTY

While living trellises may not at first seem like a traditional companion planting strategy, they are indeed just that, with one plant providing a benefit for another in the form of structural support. However, they likely don't improve the soil, deter pests, manage diseases, or meet any of the other specific goals of companion planting outlined in chapter 1. What they *do* do is look beautiful. So I'm confessing right out the gate that, in this chapter, I'm breaking one of my own rules.

Unlike the rest of the companion planting strategies presented in this book, those discussed in this chapter aren't here because they're backed by a study or careful research. Instead, they're here because they are practical techniques that work. Though "I use it in my garden and it works great" doesn't fly when it comes to maintaining a focus on research-backed plant partnerships, in this chapter, I'm giving it a pass. And, yes, I use living trellis plant partnerships in my own garden, including several of the ones described here, and they work great. However, I'm making no claim that the partnered plants provide any sort of benefit to each other except by one providing a structure for their partner to climb and making the garden look more attractive. That's as deep as these partnerships are intended to go.

The examples in this chapter match the physical structure of a tall, sturdy plant with a vine crop. Possible side benefits of these partnerships may include things like improved pollination rates if a flowering plant is used as the living trellis, reduced disease pressure if the vine crops are kept off the soil, or even enhanced biological control if one of the partners is particularly attractive to pest-eating beneficial insects. Utilizing living trellises also improves yields by taking advantage of vertical space and layering plants together. There's little doubt that after you've tried a few of these combinations and seen how easy they make it to grow more edibles in less space, you'll experiment with incorporating more living trellis plant partnerships into your garden.

PLANT PARTNERSHIPS *as* LIVING TRELLISES

For living trellises to work, two plant species are planted together based on their individual growth habits. One crop is selected for its upright, sturdy structure, and the other is selected for its vining growth habit (think of the corn and beans in the traditional "three sisters" method). Partner the two side by side, and you have yourself a living trellis.

Typically, living trellises are created for a single season, though you could certainly use a woody plant, like a tree or a shrub, to support a vine crop. However, this can sometimes have negative effects on the tree or shrub, including broken branches, reduced photosynthesis, increased competition, and the like. In many ways, the success or failure of combinations that use woody plants depends on the vigor of the trellis plant and how it's able to cope with the possible stress of being "invaded" by another plant.

With annuals, though, the plant partnerships are temporary. And if their planting is timed properly, the trellis plant is given a head start over the vine crop, allowing it to be always a step ahead. The trick is to match a relatively fast-growing, vigorous support plant with a vine crop that readily clings to it without growing too heavy for its partner to bear. In some cases, you'll have to do a bit of training to make sure the vine crops know where to go for support, but in many cases, they'll find their way up their companion plants with little or no help from the gardener. Ideally, both plant partners should provide a little something to you, too, whether it's harvestable fruits and vegetables, grains, flowers for cutting, or nectar for pollinators.

When it comes to the partner that does the climbing, I've found the most success when using crops that produce smaller fruits. Full-sized watermelons, squash, and pumpkins are likely too heavy. For this reason, the plant combinations featured in this chapter focus on vine crop varieties that are highly productive, yielding prolific harvests of sized-down veggies that won't overwhelm their plant partners.

Here are 12 excellent living trellis plant partnerships. Some of the support plants are edible, though not all of them are. Those that aren't edible are certainly ornamental, providing flowers or foliage worth admiring. Plus, almost all of these partners are easy to swap because, unlike the other plant partnerships in this book, these combinations are based on aesthetics and physical compatibility alone. Feel free to experiment, but be sure to take into consideration the size and growth habits of each plant.

Amaranth serves as a living trellis for vining plants such as small gourds.

Corn + Pole Beans

Yep, the classic combination of corn (*Zea mays*) and pole beans (*Phaseolus vulgaris*) works. The trick, though, is to give the corn a head start. Plant the corn several weeks in advance of the beans, and allow it to grow 4 or 5 inches tall before planting the bean seeds. If the two crops are planted at the same time, the beans can quickly "drown out" the corn seedlings. This plant partnership works with sweet corn, popcorn, and ornamental corn. You'll want to stick with corn varieties that grow to a height of at least 5 feet; dwarf corn selections may not hold up under the weight of the bean vines.

Corn can be planted by direct seeding as soon as the danger of frost has passed, but if you'd like your corn plant to have a bigger head start, or if you live in a gardening zone with a short growing season, sow the corn seeds in plantable peat pots indoors under grow lights four to six weeks before transplanting them outside. Because corn is a wind-pollinated crop, if you want ears to be produced, grow several blocks of plants, at least 10 plants wide and 10 plants deep, for maximum pollination. To prevent accidental cross-pollination and starchy kernels, plant only one type of corn each year.

Pole beans, too, are easy to grow by directly sowing the seeds into the garden. Pole beans will only grow in an upward direction, not laterally, so partnering them with a living trellis that has tall, straight stems like corn is ideal. There are dozens of different varieties with edible pods that can be green, yellow, purple, or bicolored. Pole Romano beans, with their wide, flat pods, are another option for this particular companion planting partnership. Plant one or two bean seeds next to each corn plant, spacing the corn plants at least 8 to 10 inches apart.

Sunflowers + Mini Pumpkins

Everyone loves sunflowers (*Helianthus annuus*), but you might not have considered them for living trellis companion planting before. Tall, sturdy varieties of sunflowers are better suited for living trellis work than dwarf or slender-stemmed varieties. To support mini pumpkin vines (*Cucurbita pepo*), be sure to plant highly branched, multiflowered varieties. Tall, straight sunflower varieties are a great living trellis for pole beans, but branched sunflowers work better with pumpkins since pumpkin vines spread in many directions.

Several mini pumpkin varieties are suitable partners for sunflowers. Varieties such as 'Baby Boo', 'Lil' Pump-ke-mon', and 'Jack Be Little' produce pumpkins that are just 4 to 5 inches across. Slightly larger varieties of pumpkins may work, too, but additional support in the form of a sling for the developing fruit might be necessary if the fruits grow too heavy for the sunflowers.

For this partnership, start the sunflower seeds indoors under grow lights four to six weeks before transplanting outdoors just after the danger of frost has passed. Pumpkins are easy to start by directly sowing the seeds into the garden after the danger of frost has passed. Mini pumpkins require around 100 days to mature. For the best results, delay planting the pumpkins until the sunflowers reach a foot or two in height. Plant one mini pumpkin vine for every 5 to 6 feet of sunflower row, or two plants per 6-foot-wide circle of sunflowers.

Broomcorn + Edible Bottle Gourds

Though here in North America we tend to use bottle gourds (*Lagenaria siceraria*) ornamentally for crafting and for autumn decorations, they are edible if harvested in an immature state before the skin hardens. And they're quite delicious to boot. The flesh is used in several Asian, African, and Middle Eastern cuisines. Unlike squash and cucumbers with their yellow blooms that close at night, edible bottle gourds produce white flowers that are pollinated at night by moths and during the day by bees.

Bottle gourd vines produce long, curling tendrils that easily cling to living trellis plants such as broomcorn (*Sorghum bicolor* var. *technicum*). They require a relatively long growing season, though this factor is mitigated because you'll harvest them for eating when they're at an immature state rather wait for them to mature on the vine. They are easy to grow from seed by planting when the soil reaches at least 65°F (18°C), but to get a jump start on the season, I recommend starting the seeds indoors under lights in plantable peat pots four to six weeks before the last spring frost is expected.

Edible bottle gourd vines grow quite long, so it is best to use taller varieties of broomcorn densely planted. Plant one gourd for every 8 to 10 feet of broomcorn row. If you allow the fruits to mature, they may become too heavy for the broomcorn stalks to bear, so be sure to harvest the fruits often, when they are not more than about 4 inches long. Tipping back the vines (pinching off their growing point) early in the season leads to more side shoots and better fruit development. It also encourages the vines to spread along the broomcorn row.

As for the broomcorn half of this partnership, there's much to know. Broomcorn is a type of sorghum that's traditionally used to make broom whisks (you can still purchase brooms made from broomcorn). Unlike other types of sorghum, the seed heads produced by broomcorn are long and fibrous. The seeds are enjoyed by birds and are quite ornamental, making beautiful additions to fall decorations. Different varieties have seeds that come in colors from pink, yellow, and orange to black and red.

Full-sized broomcorn varieties grow between 6 and 15 feet tall. The stalks are woody and sturdy, making them great living trellis plants for heavier plants like bottle gourds. Dwarf varieties are less appealing for combining with bottle gourds, since they reach only 3 to 6 feet in height.

Broomcorn is easy to grow and is a suitable living trellis in almost every region of North America. The seeds should be sown 3 inches apart as soon as the danger of frost has passed. Wait until the broomcorn plants are 6 to 12 inches tall to plant the bottle gourds at their base.

Amaranth + Chayote

Amaranth lovers are already familiar with all the benefits of this amazing plant (*Amaranthus* spp.). There are some 60 species with many flower and foliage colors. Some species of amaranth produce edible greens or seeds, while others are prized for their showy spires of blooms. Several amaranth species are suitable living trellises. *A. caudatus* (also called love-lies-bleeding) produces long, drooping clusters of blooms; *A. tricolor* is another highly ornamental species with a large, elephant-head-like flower cluster; and *A. cruentus*, or Mexican grain amaranth, is grown for its tall stalks of beautiful pink flowers, though it also produces an edible grain and edible leaves. There are red-leaved varieties of this species as well. *A. hypochondriacus* is another amaranth known for its grain production.

Though any of these amaranths perform well as a living trellis, my favorite species to use is *A. cruentus*. It reaches up to 7 feet high and produces blooms for many weeks. If you decide to harvest the seeds, they can be ground into flour or popped over heat like popcorn. Even the leaves are edible and can be cooked much like any garden green.

Amaranth seeds are tiny, and it's hard to believe they'll grow into a 7-foot-tall plant in just a few months. But thanks to their fast-growing nature and sturdy stems, these plants are a boon to your gardening self-esteem. Sow the seeds in early spring, three weeks before the danger of frost has passed. Because the seeds are so tiny, it's difficult to space them properly at planting time. Instead, thin the seedlings to a spacing of 6 to 8 inches when they're a few inches tall. You can plant them in a row or create a circle of plants to support your vines. Be forewarned, though, that amaranth sometimes self-sows quite prolifically. If you don't want a plethora of amaranth seedlings popping up in the garden the next season, be sure to trim off the flower heads before they drop seed.

Chayote (*Sechium edule*) may not yet be on the list of foods you grow, but don't let that stop you. This edible gourd is native to Central America. It's cooked like a summer squash, though the roots, leaves, stems, and underground tubers are edible, too. The pear-shaped fruits have bumpy, glossy green skin and white flesh. The seeds inside the fruits are flat, like a mango seed, and the leaves are heart shaped. Like other squashes, chayote produces separate male and female flowers on the same plant. The male flowers are produced in multiples, while the female flowers are single.

In Creole cuisine, chayote is called *mirliton*, but this veggie is used in many different cuisines all around the world. As a warm-season tender perennial, chayote grows best in warm or hot regions, like the southern United States, California, and the Gulf states, but gardeners in other regions may find success as well. Because it requires 120 to 150 growing days before harvest, it's best to start a new chayote plant from a tuber rather than from seed. Though the tubers can be difficult to find in the trade, if you know someone who grows this plant, ask him or her to dig one up and share a piece of tuber for planting. Chayote seeds germinate well only inside the rotting fruit, so if you want to grow this veggie from seed, you'll need to plant an entire fruit into the ground, not just the seed.

Chayote vines are quite prolific. I recommend one vine for every 15 to 20 feet of amaranth plants or one plant per 8-foot-wide circle of amaranth stalks. A single vine can bear dozens of fruits over the course of the growing season.

The underground tubers of chayote vines are easy to overwinter in areas where winters are cold. Simply dig them up and store them like dahlia tubers in a box of peat moss.

Quinoa + Bitter Melons

An annual grown for its edible seeds, quinoa (*Chenopodium quinoa*) is in the amaranth family and is closely related to spinach. Originating in South America, this statuesque plant has woody stems that can be branched or not, depending on the variety. Even if you do not plan to harvest the seeds, quinoa plants are beautiful ornamentals. The long spires of flowers are green or pink, and the seeds are small and black, red, or white, depending on the cultivar. Quinoa seeds are coated with bitter compounds called saponins, making them inedible unless they're treated prior to consumption. If you grow quinoa as an edible living trellis, after the harvested seeds are fully dried and threshed, wash the seeds repeatedly with fresh water to remove the bitter saponins before drying them again for storage. Washing should take place until no soaplike bubbles appear in the wash water.

If you live in a northern growing zone, look for quinoa varieties that are short season and require a shorter maturation period. To grow quinoa, sow seeds in any well-drained area with average garden soil. Know that you'll need a lot of plants if you are looking to get a sizable harvest.

On a backyard-garden scale, you can plant quinoa seeds directly into the garden around the time of your last spring frost. The best seed-sowing temperatures are between 65° and 77°F (18° to 25°C). Quinoa is not as fast growing as some other crops, so wait to sow your bitter melons until the quinoa is 8 to 10 inches tall. Thin the quinoa seedlings to 14 inches apart, as the plants will grow large and reach 8 feet tall. For gardeners in northern growing zones, get a jump start on the season by starting quinoa seeds indoors under grow lights six to eight weeks before the danger of frost has passed. Then move the seedlings outdoors.

Bitter melons (*Momordica charantia*) are beautiful plants. The leaves are deeply lobed, and the spiky fruits are incredibly unique looking. As with members of the cucumber family, the male and female flowers are borne separately on the same plant; only the female flowers will produce the bumpy fruits. A native of Asia, the bitter melon's fruits are elongated and covered in wartlike bumps and ridges and are best harvested at less than 8 inches long. The vines grow 10 to 15 feet in length, producing curled tendrils that easily grasp climbing structures or living trellises. As the vines lengthen, pinch out the growing point every few weeks to encourage more lateral branching and heavier yields.

Though the taste of raw bitter melons is bitter enough to cause your mouth to pucker, it certainly grows on you, and proper cooking mellows the bitter flavor. Best harvested when light green, the fruits should have their seeds and spongy pith removed prior to cooking. There's no need to peel them, but salting them before cooking can remove more of their bitterness. After salting, let them sit for 15 minutes, then rinse and prepare. These cucumber-like melons can be pickled, fried, or served in soups, stir-fries, or curries. They're also great stuffed with meats and grains.

Called *karela* in Indian cuisine, bitter melons are easy to start from seeds and are tolerant of many soil types. Sow seeds about a foot apart along the row of quinoa. Or, if you grow a circle of quinoa, plant two or three vines for every 6-foot-wide circle.

Sunchokes + Cucamelons

Sunchoke, also known as Jerusalem artichoke (*Helianthus tuberosus*), is a North American native perennial prized for its nutty-flavored, underground tubers with a potato-like texture. A prolific crop with stalks that reach up to 10 feet high, sunchoke plants make great supports for a sweet little cucumber-like vegetable known as the cucamelon. Unfortunately, sunchokes have weaker stems than some other trellis plants discussed here, which means they aren't good supports for larger-fruited vines or aggressive growers, like mini pumpkins, gherkin cucumbers, or bottle gourds. Because sunchokes themselves can spread aggressively, consider keeping your sunchoke crop out of the vegetable garden. Instead, plant it in an isolated patch where overzealous growth can be easily controlled by mowing around the perimeter on a regular basis.

There are several varieties of sunchokes, each of which bears tubers with a slightly different color and shape. The plants produce beautiful, multibranched clusters of 2-inch-wide yellow flowers in late summer. The knobby tubers of sunchokes are harvested anytime from late fall throughout winter (as long as the ground remains unfrozen), but the flavor is choice after the plants have been exposed to a few hard frosts. Harvest only the largest tubers and leave the smaller ones in place for subsequent harvests each fall. Sunchoke plants are winter hardy down to –30°F (–34°C).

Cucamelons, also known as Mexican sour gherkins (*Melothria scabra*), look like Barbie doll–sized watermelons. Their flavor is often described as cucumber with a hint of citrus and melon. The grape-like fruits are produced prolifically on narrow but wide-spreading vines covered with tendrils that easily grip sunchoke stems or other support structures. They are grown in a similar fashion to cucumbers, but the seeds are far smaller. Plant cucamelon seeds in spring, as soon as the danger of frost has passed. The vines may take a few weeks to get going, but once they do, they spread fairly rapidly.

Cucamelons easily reseed from any fallen fruits left behind in fall, returning to the garden year after year. They also produce an underground tuber from which the vine resprouts the following year in milder climates. If you live in a gardening zone that regularly dips below freezing, lift the tuber from the ground and store it in a box of peat moss in the garage for the winter and replant it come spring.

For this companion planting technique, plant whole sunchoke tubers in early spring and allow the plants to grow for a full year before sowing cucamelon seeds alongside the growing sunchokes in subsequent years. Once your sunchoke patch is established, it's easy to sow the cucamelon seeds around its periphery.

Sunchoke stems make a great living trellis for delicate cucamelon vines.

Orach + Fall Peas

Though we often think of peas (*Pisum sativum*) as a spring crop, they thrive in the cool temperatures of autumn. This is a beautiful, late-season companion plant partnership. A brightly colored plant with edible leaves, orach (*Atriplex hortensis*) makes an excellent living trellis for fall-planted peas. Also known as mountain spinach, orach is not only sturdy and highly branched, it's also downright beautiful. Reaching 5 feet tall at maturity, it has a mild, nutty flavor when the leaves are used raw in salads or cooked like spinach. The foliage can be green, yellow, red, purple, or magenta, depending on the variety.

Orach plants reach maturity by mid to late summer, but the greens can be picked from even very young plants and are heat tolerant. Harvest individual young leaves throughout spring and early summer, but leave the growing points intact. Pinch off the seed heads before they dry and drop seed.

Garden peas for this companion planting strategy fall into one of three categories: snow peas, with flat, edible pods; sugar snap peas, with plump, edible pods; and shell peas, with inedible pods. Any of these types of peas are suitable for planting with orach. Within each category of pea, there are dozens of different varieties. Be sure to select one with a maximum height of 4 feet to ensure the pea plants don't outgrow their living trellis.

Plant orach seeds directly into the garden two to three weeks before the last expected spring frost and allow them to grow through spring and early summer. Then, when late summer arrives, plant the peas for an autumn harvest around the base of each orach plant. Peas are easily grown from seed planted directly in the garden, but since they're cool-season crops, timing is everything with this combination. For this plant partnership to work, plant three pea seeds at the base of each mature orach for a late-season pea harvest. The pea tendrils easily grip onto the orach stalk and climb up into the branches, making a bright and beautiful combination.

Tithonia + Malabar Spinach

A relative of sunflowers, tithonia (*Tithonia rotundifolia*) is also known as the Mexican sunflower. This tall, highly branched annual makes a stunning garden specimen. With a 3-inch-diameter "trunk" and many lateral branches, tithonia serves as a great living trellis. The 2-inch-wide orange or yellow flowers are adored by butterflies and other pollinators. Tithonia transplants are sometimes available from nurseries at spring planting time, but it's easy to start the plants from seed sown indoors under grow lights. If you live in a region with a long growing season, the seeds can be sown directly into the spring garden. If you're growing them indoors to get a jump start on the season, plan to sow the seeds about six weeks before your last expected spring frost and move the transplants out into the garden when frost no longer threatens.

Since tithonia is slow to start in spring, the ideal companion planting partner for it is another slow-to-start plant, a vining crop known as Malabar spinach (*Basella rubra*). Malabar spinach is a heat-tolerant green that is downright gorgeous. The heart-shaped leaves are thick, succulent, and deep green, and the twining stems are hot pink. The leaves can be used in salads, omelets, stir-fries, or any other recipe that calls for spinach. Though it's most often grown as an annual, in frost-free areas, it's a perennial crop. Malabar spinach is very frost sensitive, so don't plant it until well after the last spring frost. It won't take off until the temperatures soar, just like its living trellis partner, tithonia.

Seeds of Malabar spinach can be sown indoors under grow lights about six weeks before your last frost for an earlier start on the growing season. But in areas with at least 100 frost-free days, you can grow it from seeds sown directly into the garden. Nicking or scraping the seed coat with a knife or sandpaper speeds germination.

Plant one Malabar spinach plant for each tithonia, and time the planting so the tithonia is 6 to 12 inches tall when the spinach is planted.

Kiss-Me-Over-the-Garden-Gate + Gherkin Cucumbers

Few annual plants make the same impact as kiss-me-over-the-garden-gate (*Polygonum orientale*). Reaching a height of up to 8 feet, this stunner boasts a thick, sturdy stem and many side branches, each covered with drooping clusters of bright pink flowers. Kiss-me-over-the-garden-gate grows so fast that you can almost watch the stems elongate right before your eyes. It makes a wonderful cut flower that's best used as a filler in bouquets, and several small native bee species enjoy drinking nectar from this Asian plant's blooms. Because kiss-me-over-the-garden-gate self-sows, once the plant is established in your garden, it returns year after year. I've never found it to be invasive in my garden, but should it become invasive in yours, care should be taken to remove as many of the flowers as possible before they drop seed.

Easy to grow from seeds sown directly into the garden or started indoors under grow lights, kiss-me-over-the-garden-gate is tolerant of early spring frosts, especially when the plants have self-sown, though I wouldn't recommend putting out transplants until the danger of frost has passed.

Small gherkin cucumbers (*Cucumis sativus*) make the perfect plant partner for kiss-me-over-the-garden-gate. The tiny, tasty fruits of these little cucumbers won't weigh down the branches too much and offer a delicious harvest for months during summer and early fall. 'Parisian Pickling', 'Vert Petit de Paris', or the West Indian burr gherkin (*C. anguria*) are a few petite cucumbers worth trying with this plant partnership. But if you're willing to harvest the fruits when they're very young, almost any pickling cucumber could be grown in combination with kiss-me-over-the-garden-gate. The cucumber vine's tendrils easily grip their living trellis to guide the plants in an upward direction, where their miniature cucumbers will be easy to harvest.

For best results, wait to plant seeds of cornichon or gherkin cucumbers until kiss-me-over-the-garden-gate seedlings are about 12 inches tall. The seeds can be sown directly into the garden. Plant one or two cucumber vines per kiss-me-over-the-garden-gate plant. You may have to train the cucumbers to grow up their trellis by aiming them in the right direction; otherwise, you may find them rambling over the soil instead.

Sorghum + Asparagus Beans

Grown primarily as a grain and sweetener (sorghum syrup or molasses), sorghum (*Sorghum bicolor*), an African native, is an important crop around the world. Varieties grown for grain are known as milo. Sorghum flowers are produced in large panicles at the tops of the stalks in late summer. Milo varieties produce tiny seeds that can be milled into flour or popped like popcorn, while cane sorghum varieties are grown for the supersweet juice that's pressed from the stalks.

Sorghum is a warm-season crop that loves the heat and, for grain usage, the seed heads are cut from the plants only after they've fully matured. Sorghum plants are straight and columnar; since the stalks can grow upwards of 8 feet tall, they're ideal partners for climbing asparagus beans. Sorghum seeds are sown two to three weeks after the danger of frost has passed. The seeds are tiny and should be spaced 8 to 10 inches apart.

Asparagus beans (*Vigna unguiculata* subsp. *sesquipedalis*) are grown for their long, edible pods. Sometimes called yardlong beans or snake beans, asparagus beans can grow as long as 24 inches, though they're best harvested when just 10 to 12 inches long. A type of cowpea, this bean tastes just like a regular garden green bean, but the flowers and beans are formed in pairs along the growing stem. Ready to harvest just 65 days after sowing, asparagus beans are easy to grow and delicious.

Once the sorghum plants reach 10 inches tall, plant one asparagus bean seed at the base of each sorghum stalk. The vines seek out something to climb soon after the seedlings emerge from the ground, so plant them as close to the base of the sorghum plants as possible. For added fun, try this partnership using the red-podded variety known as the 'Chinese Red Noodle' bean, too.

Okra + Currant Tomatoes

Standard okra varieties (*Abelmoschus esculentus*) grow between 6 and 8 feet tall, primarily in the warm southern growing conditions they prefer. Okra has hibiscus-like flowers that are followed by long, edible seedpods. Though okra is a perennial in climates where frosts don't occur, it's most often grown as an annual. The large, palm-sized leaves are borne on long petioles, or leaf stems. The plants themselves are quite beautiful, though some varieties have spines on their stalks. Excellent tall varieties for living trellises include 'Cow Horn', 'Beck's Big Buck', 'Bowling Red', 'Emerald', and 'Jing Orange'.

Okra thrives in the heat, so you may want to skip this plant partnership if you garden in the far north, where summers are cool and short. Okra pods should be picked often to encourage the plant to grow taller and produce more pods. Spined varieties make harvesting companion crops difficult, so 'Clemson Spineless' is a good, tall choice if you don't want to deal with the spines. The leaves on okra can also irritate the skin; wear long sleeves for harvest.

Okra is best grown from transplants either purchased from a commercial nursery or grown indoors under lights, and care should be taken to select a short-season variety if you want to grow this vegetable north of the Mason-Dixon Line; otherwise the plants may not grow tall enough or produce edible pods. Since okra much prefers warm weather, delay planting until the threat of frost has long passed.

Tiny, currant-sized tomatoes (*Solanum lycopersicum*) are the perfect companion planting partner for standard okra plants. Smaller than cherry tomatoes, these varieties are indeterminate so they produce long, spindly stems, and the extrasmall fruits mean they won't weigh down the okra stems. Tomatoes do not produce tendrils or wrap themselves around climbing structures, which means tomato vines must be trained to grow up into the okra plant by tying them there every few feet along their stems. Currant tomato varieties are very productive plants with sweet, thumbnail-sized fruits. 'Matt's Wild Cherry', 'Hawaiian Currant', 'Sweet Pea', and 'Mexico Midget' are good choices.

Grow tomatoes by purchasing transplants from a nursery or starting the seeds indoors under grow lights six to eight weeks before the last expected spring frost. Move the plants outdoors only after the air and soil have warmed. To use currant tomatoes in partnership with okra, wait until the okra plants have reached 2 feet tall before making a late planting of the currant tomatoes. Partner one okra plant with a single currant tomato plant.

Tree Kale + Runner Beans

Also called tree collards or walking stick kale, tree kale (*Brassica oleracea* var. *acephala*) is one unique plant. Best suited to regions with milder climates, tree kale is winter hardy down to about 20°F (–7°C). In colder areas, it will die back to the ground, negating its ability to be used as a living trellis. But in warm climates, tree kale grows 8 feet tall over the course of several years. Unlike other types of kale, tree kale does not typically flower or set seeds, and it's most often purchased as a cutting or a live plant. As the plants age, their "trunks" grow woody, and the edible leaves are found only at the very top of the plant. They look a bit like Dr. Seuss trees, to be honest, but the leaves are crunchy and delicious, and they can be used in the kitchen just like kale or collard greens. For added interest, try growing purple-stemmed varieties, too.

Once you have a colony of several tree kale plants established, they can be used as a living trellis for runner beans (*Phaseolus coccineus*) or any other type of climbing beans. Because tree kale stems are typically bare, it's easy for the bean plants to coil around them and grow. One bean plant per tree kale stalk is best.

Runner beans are often grown as ornamentals here in North America, but they are prized edibles in many parts of the world. Runner beans produce long, flat pods even in cool weather, though the pods aren't usually formed until fairly late in the season. Runner beans most often produce red flowers, but there are varieties with pink, white, or bicolored flowers. They are true perennials, though we grow them as annuals. In Hardiness Zones 8 to 10, runner beans overwinter via their fleshy, tuberous roots. In cooler climates, those roots can be dug up in fall and overwintered in the garage in a box of peat moss.

Going from seed to harvest in about 80 days, runner beans will not tolerate frosts. Care should be taken to plant them only after the danger of frost has passed. They're easy to grow from seeds sown directly into the garden. The pods can be harvested and eaten fresh, but they should be cooked prior to consumption. Runner beans, like many bean varieties, contain a natural toxin called phytohemagglutinin when uncooked. Because of this, it's important to soak and cook beans thoroughly before consuming large amounts of them.

CHAPTER 5

PEST MANAGEMENT

Luring, Trapping, Tricking, and Deterring Pest Insects

Reducing pests tends to be the most common goal of companion planting. Concerns about the possible effects pesticide chemicals can have on untargeted living organisms — including humans, pets, and pollinators — have many of us turning to more eco-friendly ways to manage undesirable bugs. While it helps to be more tolerant of pest damage to begin with, an alternative is to create a diverse, mixed habitat within the vegetable garden to limit the pest outbreaks that occur more frequently in monocultures.

A potato leaf infested with Colorado potato beetle larvae

A larval Mexican bean beetle munches on the leaves of a pole bean plant.

STRATEGIES *for* PEST CONTROL

You've already learned a bit about how the polycultures created through companion planting can influence plant health by improving soil and controlling weeds, and about how viewing the garden as an ecological habitat with many layers of interaction encourages healthier, more resilient plants. In this chapter, we'll take a look at how these same two things impact pest numbers in very practical ways.

Though gardeners have been practicing companion planting for pest control for generations, the mechanisms behind these interactions haven't been well understood. In fact, there are still many more questions than answers when it comes to how and why the presence of certain plants may (or may not) influence pest numbers and the amount of damage. While old wives' tales persist, there are also many companion planting strategies for pest management that are now backed by solid scientific research.

This chapter will introduce you to the five main categories of companion planting for pest management:

+ Luring pests away from plants through trap cropping
+ Disrupting insect feeding behaviors through masking/hiding host plants
+ Interfering with pest egg-laying behaviors
+ Using plants to physically impede the movement of pests toward a crop
+ Employing a general polyculture approach to improve overall diversity

But before we examine each of these five strategies and outline dozens of actionable plant partnerships that fit under their umbrellas, it's important to understand how pests find host plants in the first place and how companion planting can play a role in how readily a pest finds its lunch.

Cabbage loopers feed on many cole crops.

Whiteflies feed on a broad range of plants, including tomatoes.

How Do Pests Find Their Host Plants?

Plant-eating insect pests locate their host plants via a multitude of cues. In most cases, it's a combination of visual cues based on the appearance, color, and size of the host plant, combined with a series of chemical cues, including the detection of volatile chemicals (similar to odors) released by the host plant. (There's a third possible kind of cue as well, but I'll get to that one a bit later — see page 98.) When the insect is some distance from its host, the volatile chemicals may be their biggest beacon, but as the insect approaches the plant, visual cues likely become more important.

Because both sets of cues influence how a pest finds its host, disturbing one or both of those cues in some way can be a key to keeping pests off plants. For example, if the pest can't recognize the volatile chemicals being released by the host plant because the signal is being masked by the presence of another volatile chemical, the pest might have a more difficult time finding a place to feed. Or if the visual cues are somehow masked such that a pest can't recognize its host, that, too, could make it hard for the pest to find a suitable host plant. In theory, by planting companion plants that "confuse" pests, gardeners can disrupt a pest's host-seeking behavior, leading to a reduction in the number of pests and therefore the amount of pest damage.

The truth is, however, that there are so many factors in play, this masking is often easier said than done. Each pest species acts differently, and there's great variation in how strongly drawn each pest is to a particular host plant — not to mention the fact that many pests have multiple possible host plants. How fast and far any given pest can move is another influencing factor, and some of the smaller, more generalist pests (like aphids and whiteflies) that feed on a broad range of plant species locate their host plants randomly rather than through a set of cues. One study found that larger insects tended to be easier to "throw off the scent" of their host plant, while tiny insects were less deterred.

Masking versus Repelling

As you might surmise, most companion planting techniques aimed at limiting pest damage do so with the goal of disrupting host-seeking behaviors, regardless of whether the pest is going to eat the plant directly or lay eggs on it. The presence of certain volatile chemicals from a companion plant can mask the volatiles being released by a particular pest's host plant, and if the companion plants are interplanted tightly, they may visually confuse the pest as well by making the host plant less distinguishable — hiding it, in a sense.

It's important to note that this masking effect is different from what was believed for many decades. It was once thought that the odor of certain companion plants deterred pests from feeding. For example, commonly recommended companion plants, such as marigold, peppermint, sage, and thyme, were often chosen for their strong fragrance. But over many years, multiple studies have shown that the scents of these plants don't necessarily *repel* pests. Instead, further data are suggesting that these odors *mask* the volatile chemicals emitted from host plants. This points to the fact that the mechanisms of how certain companion plants deter pests (or don't, as the case may be) are still being researched and debated among scientists.

However, plenty of pest-deterring strategies are backed by solid scientific research, and those are the ones we will consider here. The five broad categories of companion planting for pest management that I touched on earlier all take the ways herbivorous insects find their host plants and turn them to the gardener's benefit to help manage pests. Now let's look at each of these categories in depth and lay out specific plant partnerships that have been shown to be effective for each.

A Word of Caution

Despite all the upsides of companion planting for pest management, there are a few downsides, too, and competition for resources is chief among them. If not carefully selected and managed, the companion plant may overtake the vegetable crop and cause decreased yields. When following the strategies in this chapter, it's important to be sure your desired crops still have plenty of access to light, water, and mineral nutrition. Slower-growing vegetable crops may be more readily "swallowed" by neighboring companion plants than quicker-growing varieties. This must be factored in when determining how many companion plants to include, how closely to plant them, and how to time the plantings.

The Appropriate/Inappropriate Landings Theory

An additional theory about how and why mixed plantings deter pests has to do with another way scientists think certain pest insects may find their hosts: through direct contact. It's called the appropriate/inappropriate landings theory.

Some fascinating research has pointed to the fact that plant-munching pests may locate host plants not just through visual cues and volatile chemical signals but also by making a series of landings on the plant's foliage and "tasting" it with receptors on their feet. Researchers observed flying pests making a series of repeated landings on a host plant before settling down to lay eggs. They noted that the pests needed to make a specific number of appropriate landings on the foliage of their target plant before they received enough stimuli to initiate egg-laying behavior. Then the scientists discovered that when the host plants were interplanted with a companion plant, the pests ended up making an occasional landing on the nontarget plant, throwing off the number of appropriate landings and making it less likely that the pest insect would settle down and lay eggs. The researchers noted that 36 percent of the study insects being observed laid eggs on host plants that were grown in bare soil, while only 7 percent laid eggs on host plants that were surrounded by companion plants.

This theory, if proven true in repeated studies, could have a significant impact on how we garden. When crops are grown in a monoculture, the chance of a pest making the required number of appropriate landings is much greater than in plots with mixed plantings. This also would explain why certain insects are pests in agricultural settings but not in a mixed natural habitat.

When crops are grown in bare soil or in a monoculture, the chances of a pest finding them are much greater than they are for crops grown in mixed plantings.

PLANT PARTNERS *for* TRAP CROPPING

Trap cropping is a companion planting technique for pest management where veggies are planted in conjunction with "sacrificial" companion plants whose economic benefit comes solely from their ability to draw pests away from the vegetables to be harvested. Selecting a companion plant because it's more attractive to pests than the desired crop is what this process is all about.

Trap cropping has been used by farmers for centuries to shield crops from insects. On a commercial scale, some farmers may employ trap crops not just to draw pests away from desired vegetables but to concentrate them into a smaller area so they can target pesticide sprays to the trap crop and not have to apply them to the edible crop (a process sometimes called attract-annihilate). While home gardeners can certainly use this practice, too, in many instances the mere presence of the correct trap crop nearby will be enough to diminish pest damage to a tolerable level without having to use pesticides.

In essence, trap cropping manipulates pest behavior by luring pests to a new location. Since most insects show a preference for some plant species over others, and even for plants in different developmental stages, timing and placement are essential in trap cropping, and it's important to choose the best partner plants to ensure the highest rates of success. Sometimes a single trap crop is used, while other times it's a blend of multiple plants that will generate the most success (see Building a Push-Pull System, page 108).

Typically, trap crops are planted a few weeks in advance of the vegetable crop so the plants are able to lure in the pests early in the veggie crop's growing cycle. Trap cropping has been shown to be effective against a number of different pests, but it's particularly useful against insects that locate their host plants through a combination of visual and volatile chemical cues, rather than those that find their host plants by accident (aphids or mites, for example, who often arrive randomly on a wind current).

When employing trap cropping, the design and arrangement of the vegetable crop and the trap crop depends on the behavior and mobility of the insect, the site itself, and other factors. For example, in most cases, planting the trap crop on the periphery of the garden draws the pest away, but in some cases, especially with pests that aren't as mobile, interplanting the vegetable with the trap crop may be more effective. It does require a bit of experimentation to see which placement works the best, so be prepared to try different arrangements and take some notes.

That said, here are some general guidelines for home gardeners to follow regarding how to arrange and implement a trap-cropping companion planting system:

+ Locate the trap crop on the perimeter of the garden, several feet from the desired crop, if the pest is highly mobile (Colorado potato beetles, squash bugs, harlequin bugs, or lygus bugs, for example) or if the pest is the offspring of a highly mobile insect (cabbageworms or diamondback moth caterpillars, for instance).
+ Interplant the trap crop very close to the vegetable crop, or in alternate rows, for pests with limited mobility (aphids, mites, whiteflies, and flea beetles, for example).

Here, a trap crop of radish is planted around a young tomato plant. Flea beetles are drawn to the radish leaves, minimizing damage to the tomato transplant.

- The size of the trap-crop planting should be based on how problematic the pest is and how mobile it is. In general, the size of the trap-crop planting area should be 10 to 20 percent of the area of the vegetable crop.
- Time the planting of the trap crop such that the trap plants are several weeks more mature than the vegetable crops.
- If you don't want to employ pesticides to destroy pests collected on trap crops, but you're still looking for a way to get rid of the little buggers, invest in a wet/dry vac or use an old vacuum cleaner to suck the pests off your plants. Works like a charm.
- Another effective strategy is to place the trap crop between the garden and a wild space where pests may have overwintered. This gives the pests a more convenient food source and keeps them from migrating over to the main crop

In agricultural systems, trap cropping is as much art as it is science, involving many different strategies. Farmers often use more nuanced trap-cropping plans than we'll touch on here. For example, control of stink bugs and leaf-footed bugs was obtained in a Florida study by combining buckwheat, millet, sorghum, and sunflowers around the perimeter of a vegetable crop. However, most home gardeners would not have the space or resources to grow such an extensive mix of trap crops. Instead, basic single-species trap cropping is likely the best option for home gardeners. Thankfully, there are plenty of simple, easy-to-employ strategies perfectly suited to home gardens.

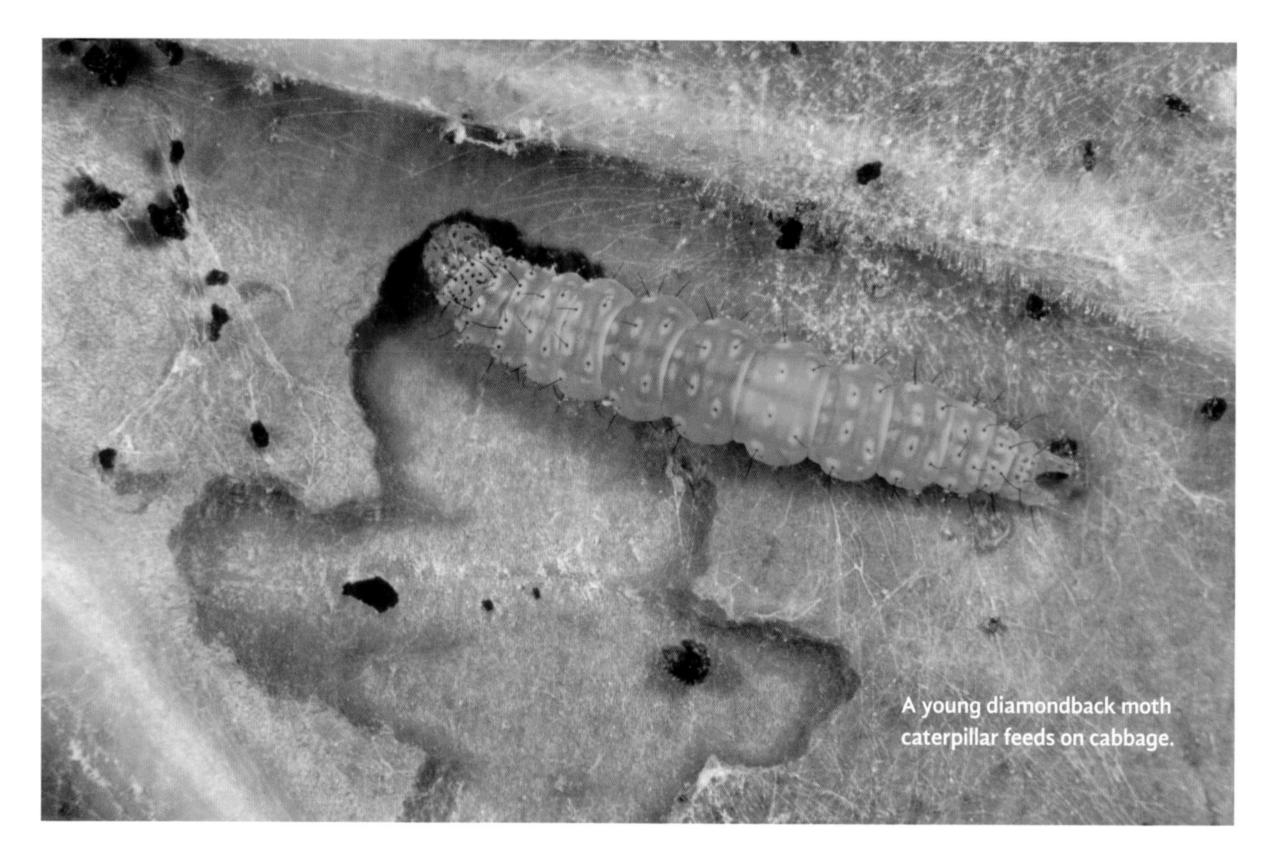

A young diamondback moth caterpillar feeds on cabbage.

Cabbage + Collards to Lure Diamondback Moths

Collards and cabbage are both members of the cole crop family, and both are susceptible to damage from diamondback moth larvae. However, adult diamondback moths prefer collards over cabbage for egg laying. Therefore, planting collards several feet away from the cabbage crop lures the adult moths away and limits damage to cabbage plants.

Diamondback moths are small, nondescript, night-flying moths that lay eggs on host plants. The eggs hatch into small caterpillars that wiggle and drop off plants when disturbed. They feed mostly on the undersides of leaves and flower stalks. Diamondback moth caterpillars feed on a wide array of cole crops, so a trap crop of collards may protect broccoli, cauliflower, kale, and other cole crops from this pest as well.

Green stink bugs enjoy feeding on tomatoes, but they like cowpeas even more.

Tomatoes + Cowpeas to Lure Southern Green Stink Bugs

Cowpeas were introduced in chapter 2 for their usefulness as a warm-season cover crop. However, they've also been found to be an excellent trap crop for southern stink bugs.

Southern green stink bugs are an introduced species of insect that has become a pest in the southern United States. The adults are green and shield-shaped. They're fairly large (around ½ inch long), and females can lay nearly three hundred eggs in their lifetime. Southern green stink bugs feed on tomatoes, peaches, beans, and many other plants. They have needlelike mouth parts they use to inject saliva into developing fruits and vegetables. They then suck the juices of the dissolved plant tissue, causing stippling and corking on the fruit. Heavily damaged fruits may drop off the plant before they develop.

Southern green stink bugs are good fliers, so for this trap-cropping partnership to work, plant cowpeas several feet away from tomatoes or other susceptible crops. Be sure to plant the cowpeas several weeks before planting the crop to be harvested. For best results, collect and destroy any stinkbugs you find on the cowpeas. This further limits stinkbugs in your desired crops.

A larval squash vine borer inside a cut-open squash vine

Various Squash + Blue Hubbard to Lure Squash Bugs and Squash Vine Borers

Blue Hubbard squash is an old-fashioned variety of squash that produces big, hard-skinned fruits with a bumpy, bluish green rind. Each large squash is round in the middle and tapered on the ends. While blue Hubbard is an excellent eating squash, it's also an exceptional trap crop for both squash bugs and squash vine borers, since they much prefer it to other squash varieties. Research has shown that when blue Hubbard is planted on the periphery of a patch of squash types, the insects are drawn toward it and away from the other varieties.

Squash bugs are common pests of melons, pumpkins, squash, and zucchini. They're brown with an elongated shield shape. Both adults and nymphs use their needlelike mouthpart to suck juices from plants and developing fruits. High infestations cause the vines to wilt and turn brown.

Adult squash vine borers are red-and-black, clear-winged moths that fly during the day. Females lay eggs at the base of squash, melon, and pumpkin plants. The resulting grublike larvae tunnel into the plant stem, consuming the tissue inside and eventually killing the plant.

When using blue Hubbard squash as a trap crop for squash bugs and/or squash vine borers, be sure to plant it three to four weeks prior to planting other squash varieties. Locate blue Hubbard on the periphery of the vegetable garden, or at least several feet away from other squash varieties.

Cherry peppers are favored hosts for egg laying by adult pepper maggot flies.

Bell Peppers + Hot Cherry Peppers to Lure Pepper Maggots

Researchers found that using a trap crop of hot cherry peppers around the perimeter of bell pepper plantings reduced pepper maggot infestations significantly. Farmers testing this trap-crop method in Connecticut reported a 98 to 100 percent reduction in pepper maggot flies in their bell pepper crops.

Pepper maggot flies are common all up and down the East Coast and as far west as Texas. Adult pepper maggot flies lay eggs directly on peppers. The maggot then tunnels into the fruit and eats the tissue inside. When it's ready to pupate, it chews a hole in the bottom of the pepper and drops to the ground. Most pepper maggots aren't found until the peppers are cut open or the fruit rots prematurely on the plant. Some varieties of peppers are preferred by pepper maggot flies for egg laying, including hot cherry peppers, making them a great trap crop.

For this trap crop plan to work, plant hot cherry peppers around the periphery of your pepper patch, or plant a row on the outer edge of the garden. The adult pepper maggot flies will lay their eggs on the hot cherry peppers rather than on your sweet peppers.

Pak choi with flea beetle damage grows around an undamaged pepper plant.

Various Vegetables + Radish or Pak Choi to Lure Flea Beetles

Flea beetles are a notoriously challenging pest for vegetable gardeners. They are tiny members of the beetle family, measuring only about 1⁄10 inch long, and they hop like fleas. There are about 20 different species of them in North America. Adult flea beetles are chewing insects that leave tiny, ragged holes behind in plant foliage. Their underground larvae feed on plant roots. Flea beetles feed on diverse crops in the garden, including eggplants, peppers, potatoes, tomatoes, and cole crops such as cabbage, broccoli, and kale. But their favorite vegetable crops include radish and pak choi (or bok choy). While mature tomato, eggplant, and broccoli plants can withstand a good bit of flea beetle damage, young transplants can really suffer. Researchers found that using radish and pak choi as trap crops reduced the amount of flea beetle damage on several different vegetables.

Flea beetles do not move very far, so to use this companion planting partnership, interplant rows of susceptible crops with radish or pak choi. Both are easy to grow by direct seeding, and the seeds are inexpensive. Sow the radish or pak choi seeds a few weeks in advance of the crop you're trying to protect.

Cole Crops + Chinese Mustard Greens to Lure Flea Beetles

Chinese mustard greens have been found to be an excellent trap crop, particularly for protecting cole crops — such as cabbage, broccoli, and cauliflower — from flea beetles.

Planted very early in the growing season, Chinese mustard attracts flea beetles and keeps them off of cole crops that are also planted early in the season. Since both the trap crop and main crop prefer similar growing conditions, namely cool weather, this combination has proven quite effective. For the best results, alternate rows of the Chinese mustard trap crop and the desired cole crops, and plant the mustard several weeks before planting the vegetables.

Various Vegetables + Mustard Greens to Lure Harlequin Bugs

Mustard greens (the variety 'Southern Giant Curled' in particular) were also shown to be among the most attractive plants to harlequin bugs. The bugs chose mustards more consistently than any other crop in field studies.

Harlequin bugs are a significant pest to members of the cole crop family, including broccoli, collards, kale, and cauliflower. They feed on many other vegetable crops, too, including beans, potatoes, squash, tomatoes, asparagus, and several different fruits and brambles. Though they're more prevalent in southern regions, the range of harlequin bugs may be expanding northward. They are large, shield-shaped insects, black with bright orange markings. Both as adults

Chinese mustard greens make an excellent trap crop for flea beetles.

A harlequin bug feeds on mustard greens.

and immature nymphs, they suck sap from leaves and fruits with a needlelike mouth part. Damaged leaves may turn brown and die. Harlequin bugs are a member of the stink bug family and will produce a foul odor when disturbed.

Since harlequin bugs produce several generations of offspring per year, they can prove extremely problematic; luring them away from susceptible plants early in the season helps limit damage and encourages them to lay eggs on the trap crop rather than on the one to be harvested.

Plant mustard greens in a row on the periphery of the garden several weeks prior to planting vegetable crops. Harlequin bugs are highly mobile, so the mustard greens should be several feet away from the crops you intend to protect.

Strawberries + Alfalfa to Lure Lygus Bugs

Lygus bugs (also called western tarnished plant bugs) attack many fruit and vegetable crops and are found across much of the United States. Adult lygus bugs range in color from pale green to dark brown, with black markings. Only about ¼ inch long, these small bugs can cause big problems for gardeners. The bugs produce three or four generations per year, and both the nymphs and the adults feed by sucking juices from growing fruits.

Lygus bugs attack plants ranging from cotton and apples to tomatoes and beans, but they're especially problematic on strawberries, causing the fruits to

Building a Push-Pull System

There's another important method of trap cropping known as a push-pull system that takes this companion planting technique and adds another layer of protection. Push-pull trapping systems combine the vegetable crop with a trap crop that lures the pests away and another plant that masks or hides the vegetable crop. In other words, the pest is "pushed" away from the host plant with a masking companion plant and then "pulled" toward a trap crop. Push-pull systems can be extremely effective, especially when the masking companion plant is interplanted with the vegetable crop and the trap crop is located on the periphery of the planting.

Alfalfa can lure lygus bus from strawberry plants.

become distorted and misshapen (called catfacing). In test plots, it was discovered that lygus bugs much prefer alfalfa plants to strawberries, and the study found that one row of alfalfa plants for every four rows of strawberries significantly reduced lygus bug damage on the strawberries. Researchers in this California experiment then used a vacuum to suck lygus bugs from the alfalfa trap crop, reducing their numbers by up to 90 percent.

In home gardens, plant a trap crop of alfalfa 2 to 3 percent the size of your strawberry patch. The alfalfa plants should be located very close to the strawberry plants, but they should not compete with the berries for the same space.

PLANT PARTNERS *for* MASKING HOST PLANTS

There is a widely known hypothesis among scientists when it comes to the ability of a pest insect to find a host plant. It's called the resource concentration hypothesis, and it proposes that plant-eating insects are less likely to find and remain on their host plant in diverse habitats. In field studies where interplanting is employed, herbivorous insects spend less time on the vegetable crops growing there. This, combined with a greater population of pest-eating beneficial insects supported by such a polyculture, leads to lower pest populations and a lower risk of pest damage.

Since companion planting is really just a way to build a polyculture, the resource concentration hypothesis likely applies to home vegetable gardens, too. You'll learn more about how companion planting can boost the numbers of beneficial insects in your garden in chapter 7. Below you'll find a list of actionable plant combinations that are all aimed at protecting host plants through masking volatile chemicals, visually confusing the insect, or limiting the number of appropriate landings the insect makes.

Peppers + Alliums for Green Peach Aphids

Green peach aphids are among the most common aphids that feed on peppers. They feed spring through fall, using their needlelike mouthparts to suck juices from the plant. Aphids are tiny, pear-shaped insects that may be winged or not. A careful look at their posterior reveals two tubes, called cornicles, sticking out on either side, which secrete alarm pheromones when the aphid is attacked by a predator. There are several hundred aphid species in North America, and most have specific host plant needs, feeding on just a single plant species or family. Aphids are most often found in groups on new plant growth or on leaf undersides. Extensive feeding can cause leaf curl and yellowing. Green peach aphids are known to transmit several plant viruses to pepper plants.

Interplanting peppers with members of the allium family, including onions, scallions, and garlic, has been shown to deter these small insects from settling onto the pepper plants to feed. Plant the selected allium companion plants in between and around pepper plants rather than just nearby.

Zucchini + Nasturtiums for Squash Bugs

Squash bugs are common pests on pumpkin, zucchini, squash, melons, and cucumbers, and they are challenging to control without the use of synthetic chemical pesticides. However, an interesting Iowa State University study found a significant decrease in the number of squash bugs and amount of squash bug damage when nasturtiums (*Tropaeolum majus*) were grown side by side with zucchini, compared to when the zucchini were growing by themselves in bare soil.

Though this particular study examined this plant partnership using only summer squash, the combination may translate well to deterring squash bugs from winter squash varieties, too. Nasturtiums are annual plants that produce lovely edible flowers, so even if this partnership isn't as effective for winter squash, the nasturtium blooms will still be a nice addition to your garden and will likely boost pollinator numbers in the squash patch.

Gardeners with limited space would do well to select a nasturtium variety with a bush-type growth habit, while those with more space may prefer one of the many vining nasturtium varieties, allowing the plants to ramble around and between the squash plants.

◀ Planting members of the onion family around pepper plants has been shown to deter green peach aphid feeding.

▶ Trailing nasturtiums growing among zucchini plants helps to deter squash bugs.

Chinese Cabbage + Green Onions for Flea Beetles

Chinese cabbage is another cole crop favored by flea beetles. When grown side by side with green bunching onions, Chinese cabbage incurred significantly less flea beetle damage. However, a Washington State University study found that green onions were not effective at deterring flea beetle damage on another cole crop, broccoli.

For that reason, intercropping Chinese cabbage or other cole crops with green bunching onions and partnering that combination with a nearby trap crop of mustard greens to create a push-pull system (see page 108) is likely to be a better method of deterring flea beetle damage on various cole crops. The green onions mask the cole crop from flea beetles while the mustard greens lure them to a different area of the garden. Since green onions, cole crops (including Chinese cabbage), and mustard greens all prefer the cool weather of spring, this push-pull system is effective from very early in the growing season. It also may help limit flea beetle damage on warm-season crops, such as eggplants and tomatoes, later in the season.

Tomatoes + Basil for Thrips

Thrips are tiny, narrow insects that feed by sucking plant juices, causing discoloration and a silvering or puckering of leaf surface, flower buds, or fruits. Thrips can spread various plant diseases. On tomatoes, they're responsible for transmitting tomato spotted wilt virus. Less than 1⁄20 inch long, thrips are a challenge to identify and control. Signs of tomato thrips include stunted growth, tiny pale spots on the fruits, and early fruit drop, along with the tiny specks of their black excrement. Often the terminal shoots of the plants are killed or stunted. Many times damage from thrips is confused with damage caused by other insects and mites.

The same species of thrips that feed on tomatoes, including western flower thrips and onion thrips, also feed on onions (they much prefer red onions to white). Because of this, if thrips are problematic in your garden, keep members of the onion family separate from tomatoes. Instead, interplant tomatoes with basil (*Ocimum basilicum*), which has been found to help mask tomato plants from thrips.

Though I wasn't able to uncover research on other plant partnerships utilizing basil to deter thrips, it makes intuitive sense to interplant with peppers, eggplants, beans, celery, potatoes, or other susceptible crops, as they, too, are common food sources for thrips. It isn't certain if basil is as effective a companion plant for those crops as it is for tomatoes, but the combination will not prove harmful.

◄ A new planting of Chinese cabbage is interspersed with young green onion seedlings to mask the cabbage plants from flea beetles.

Basil has been shown to help mask tomato plants from thrips. Plant the two species side by side to cut down on thrips damage.

Grow calendula with collards to reduce aphid numbers and damage.

Collards + Calendula for Aphids

Aphids are notorious for infesting collards and other cole crops. Combining collards with calendula plants (*Calendula officinalis*) was found to deter aphids from feeding on the collards and attract many species of beneficial insects that help control aphids (more on this in chapter 7).

Calendula, also called pot marigolds, are annuals easily grown from seed planted directly into the garden. Their edible petals are used in various recipes, as well as in crafting and herbal tea making. Single-petaled varieties may be more welcoming to pollinators and other beneficial insects than double-petaled types.

Interplanting collard greens (and other cole crops) with calendula and other flowering herbs and annuals may help reduce aphid numbers.

Potatoes + Tansy or Catmint for Colorado Potato Beetles

The common garden herbs tansy (*Tanacetum vulgare*) and catmint (*Nepeta* spp.) were both found to deter Colorado potato beetles from feeding on potato plants when the herbs were interplanted directly into the potato patch. They were less effective when planted around the periphery of the potato planting.

Tansy is a perennial flower in the aster family (Asteraceae). The yellow, button-sized blooms top green foliage that's fernlike in appearance. Tansy spreads relatively quickly by underground rootlike structures called rhizomes, but it is not considered invasive. Tansy is hardy to –30°F (–34°C). The leaves have a pungent scent that may mask the volatiles produced by potato leaves, making it difficult for Colorado potato beetles to find the potato plants.

Catmint is a great partner plant for potatoes and has been shown to deter Colorado potato beetles.

There are several species of catmint in the genus *Nepeta*. These fragrant plants are in the mint family (Lamiaceae) and produce pretty blue-purple flowers set against gray-green foliage. Most species of catmint typically planted in ornamental gardens are winter hardy to –30°F (–34°C).

The main issue with interplanting one or both of these herbs with potato plants is that the herbs are perennials, while the potatoes must be dug for harvest, disturbing the soil around the roots of the tansy and catmint. Thankfully, these two perennial herbs are tough, and if you happen to uproot them while digging out your potato crop, it's easy enough to replant them.

Another troublesome aspect of this companion planting partnership is that potatoes should be rotated to a new growing spot each season to limit soilborne diseases, which means the tansy or catmint must be moved to the new planting area along with them. Again, fortunately, these are tough perennial herbs; digging them up in spring and moving them to a new potato patch certainly disturbs them but is seldom enough to kill them, as long as they receive adequate water to reestablish in the new location.

The ferny foliage of dill, when combined with broccoli plants, has been shown to reduce the egg-laying behaviors of the imported cabbageworm.

PLANT PARTNERS *to* INTERFERE *with* EGG LAYING

The appropriate/inappropriate landings theory (see page 98) states that insects must make a set number of landings on the correct host plant to receive a strong enough cue to initiate egg-laying behavior. "Hiding" host plants from egg-laying pests through the use of a companion plant is a means to that end. The following plant partnerships have been shown to limit the egg-laying behaviors of certain pests.

Cole Crops + Sage, Dill, Chamomile, or Hyssop for Cabbageworms

Though sage (*Salvia officinalis*), dill (*Anethum graveolens*), German chamomile (*Matricaria chamomilla*), and hyssop (*Hyssopus officinalis*) are often recommended as companion plants for repelling various pests due to their fragrant foliage and high essential oil content, they haven't always proven to be effective in field studies. Several studies determined that these herbs have little to no repelling effects on insects like squash bugs and cucumber beetles.

However, in the case of cabbageworms, these fragrant plants have been found helpful at deterring the egg-laying behaviors of adult cabbageworm butterflies. When these herbs are grown singularly or in combination among members of the cole crop family, fewer cabbageworm butterflies laid eggs on the plants, either because the herbs masked the plants with their volatile chemicals or because they physically hid the host plant from the pest.

▶ In my own garden, I often plant chamomile (for tea) with my cabbage plants. The aim of the partnership is to reduce the number of cabbageworm eggs laid on the cabbage.

In addition, these herbs serve as a nectar source for various species of beneficial insects that prey upon cabbageworms and other pest caterpillars, so should any eggs happen to be laid, the herbs may lead to a reduction in the number of caterpillars that survive and the amount of damage they cause.

All of these herbs are easy to grow. Sage and hyssop are perennials with good winter hardiness, while dill and German chamomile are annuals that will return to the garden year after year if they're allowed to self-seed. Be careful, however, as both can become potentially weedy if they're allowed to drop too much seed. Harvesting some flowers for culinary use will help limit seed production, and pulling out unwanted seedlings is an easy job. Despite their tendency to self-sow a bit too enthusiastically, using the two annual herb options as companion plants may be more practical than interplanting cole crops with the two perennial types. Again, this is because cole crops should be rotated to a new site each season, making it necessary to uproot and relocate the perennial herbs along with them. I suspect most gardeners would find it far easier to simply plant new dill and/or chamomile plants at the start of each growing season in the new planting site.

Interplanting tomatoes with basil plants (red basil shown here) has been shown to limit tomato and tobacco hornworm egg laying.

A hornworm feeds on a tomato plant.

Tomatoes + Basil for Hornworms

Planting basil with tomatoes has been shown to limit egg-laying behaviors by the adult moths whose leaf-eating larvae are known as tomato and tobacco hornworms.

Tomato and tobacco hornworms are two closely related species of moths. The large brown adults are active at night, when they sip nectar from tubular flowers and lay eggs on members of the nightshade family, including tomatoes, tobacco, eggplants, and peppers. The resulting larvae start out very small but quickly grow to the size of your thumb as they munch on their host plants. Tomato and tobacco hornworms are medium green in color with white diagonal lines (tobacco hornworm) or sideways Vs (tomato hornworm) down their sides. They also have a hornlike spine protruding from their posterior.

When they're small, hornworms can be difficult to spot, especially during the day when they hang out close to the center of the plant, often beneath a stem. But when they grow large, their size makes them difficult to miss.

Planting tall varieties of basil between and around tomato plants resulted in reduced egg-laying behaviors in the adult moths. For the best results, the basil plants should be close to the tomato plants but not so close as to restrict air movement and promote fungal diseases. Try planting four or five basil plants around each tomato plant, or alternate rows of basil and tomatoes in the garden.

Tomatoes + Thyme or Basil for Yellow-Striped Armyworm

The yellow-striped armyworm is a common pest in the eastern United States and as far west as the Rockies. Some southwestern states have this pest as well. However, it's most problematic in the Southeast.

There are multiple generations of yellow-striped armyworms each season. When they first hatch, the young caterpillars feed in groups, but as they age, they separate, eventually reaching about ¾ inch in length. Adult moths are small with brown forewings and white hind wings. The adults are active at night, which is when egg laying occurs.

The larvae feed on tomato leaves when they're young, but as they mature, they also attack the developing fruits, leaving holes and tunnels behind. They eat lettuce, cabbage, beans, peppers, carrots, turnips, and many other common garden plants and flowers as well.

An Iowa State University study found that interplanting tomatoes with thyme or basil resulted in a significant reduction in egg-laying behaviors by the adult armyworm. Underplant tomatoes with a living mulch of thyme plants or surround tomato plants with basil to deter this pest. Just keep in mind that thyme is a perennial, so the plants will have to be moved when tomato plants are rotated to a new garden area each season.

Thyme is a great plant partner for tomatoes because it helps mask the plants from adult yellow-striped armyworm moths.

Yellow-striped armyworm

In this beautiful kitchen garden, a carpet of thyme is planted around young cabbage plants in hopes of reducing damage from cabbageworms and cabbage loopers.

Cabbage loopers are common pests of cole crops.

Cole Crops + Thyme for Cabbageworm and Cabbage Loopers

A study on an Iowa State research farm found that partnering broccoli and cabbage plants with thyme resulted in significantly reduced egg-laying behaviors by adult imported cabbageworm butterflies and cabbage looper moths, which are among the most ubiquitous pests of broccoli and other cole crops.

Imported cabbageworm adults are day-flying butterflies with white wings and a wingspan measuring 1 to 1½ inches across. Female butterflies lay yellow, oblong eggs on all members of the cole crop family. The resulting larvae are light green with faint, long, creamy yellow stripes down their bodies. The caterpillars can quickly skeletonize the leaves of host plants.

Cabbage looper adults are night-flying moths about the same size as cabbageworm adults. They're gray-brown in color with a white figure-eight in the middle of each forewing. Their eggs are white and shaped like a thick disc. The caterpillars are pale green with white lines down each side. Their arched, looping bodies look like inchworms when they move.

Due to a reduction in egg-laying behaviors, both types of leaf-eating caterpillars were found in lower populations in test plantings where host plants were interplanted with thyme versus sites where the host plants were grown in bare soil.

It's important to note that thyme isn't the only companion plant that can lead to a reduction in the egg-laying behaviors of these common pests. They have also been found at lower numbers when cole crops are interplanted with different types of living mulch.

Onion and Cole Crops + Marigolds for Onion Root Maggot Fly and Cabbage Root Fly

Marigolds are often recommended as a companion plant. The truth is, there isn't much science to back up many of the touted combinations. This particular companion planting partnership, however, showed positive results.

In study areas where onions were interplanted with marigold, there was a reduction in the egg-laying behaviors of onion flies. Adult onion root maggot flies are nondescript gray-brown flies. Female flies lay eggs at the base of onions and other members of the onion family (alliums), including garlic and leeks. The larvae feed on the roots, bulb, and stem of the plant. After spending the winter underground as pupae, a new generation of adult flies emerge in spring and begin to look for their host plants. Damaged areas often develop rot. Unfortunately, there can be three or more generations of onion root maggot flies in a single growing season.

A similar pest, the cabbage root fly, looks and acts much the same way but targets cabbage plants as well as other members of the cole crop family, including root crops like radishes, rutabagas, and turnips. Eggs are laid at the base of the plant, and the larval flies feed on the roots and the base of the stem. They can completely destroy the plant's entire root system. Affected plants will wilt, and the outer leaves turn yellow or purple. Eventually, the plants collapse and die. If you pull up infested plants, you may spot the live larvae feeding on the dead root system.

Cabbage root maggot fly larvae feed on the roots and the base of cabbage plants.

Grow white clover as a living mulch beneath cabbage plants to deter root maggot flies from egg laying.

Much like the onion root maggot flies, the cabbage root maggot flies showed reduced egg-laying behaviors when susceptible crops were interplanted with marigolds. Alternate rows of marigolds with cole crops and onions, or interplant the two partner plants within each row.

Cabbage + White Clover for Cabbage Root Maggot Flies

In a study done in the Netherlands, a planting of white clover around cabbage plants suppressed cabbage root maggot fly egg-laying and larval populations enough to improve cabbage quality compared to control plots. Though the cabbages grown in companionship with white clover were slightly smaller in size, probably due to increased competition, reduced pesticide costs offset the reduced weight of the cabbages.

In this partnership, cabbage rows should be underplanted with a living mulch of white clover. Because white clover is a perennial, using this companion planting partnership means you'll have to either deeply till the clover under at the end of the growing season to keep it from coming back or keep the bed of clover intact as a perennial living mulch, simply swapping out the vegetables growing in it each season to sustain a good crop-rotation program.

PLANT PARTNERS *to* IMPEDE PEST MOVEMENT

Another way gardeners can use companion planting to manage pests is by using plants to physically impede pest movement. Two basic techniques help block pests from getting to harvestable crops.

Hedgerows

The first technique is to create a barrier planting between the garden area and uncultivated areas where pests may take shelter, especially during the off-season. Hedgerows were once a common sight in rural yards and on farms, but as America has become more suburban, hedgerows have fallen out of fashion. It's time to bring them back into vogue.

Hedgerows are wide, linear, permanent plantings composed of a mixture of trees, shrubs, grasses, and perennials. The plants are often closely spaced and selected to provide a succession of blooms and a variety of textures and growth habits. Traditionally, hedgerows were planted on field edges, in orchards and vineyards, and on property lines. When properly sited, their placement may inhibit the movements of certain pests from wild spaces, where they shelter in the winter or feed when their host vegetable plants are not available, to garden areas.

Hedgerows are beneficial for many other reasons, too. They create permanent windbreaks, sheltering homes and roads from wind and blowing snow with their dense collection of branches, twigs, and stems. They also provide wildlife habitat, protect against soil erosion, filter airborne dust, eliminate maintenance-heavy turf, and increase pollinator diversity. Many species of pollinators and other beneficial insects much prefer this type of year-round shelter to that of annual and herbaceous plants.

Expansive suburban developments are perfect candidate sites for hedgerow installation. Trees, fields, and other vegetation are razed to construct these housing developments, leaving the land without windbreaks, shade, and wildlife habitat. Hedgerows can create a living fence around a property's perimeter or between the home and a road, sheltering it from noise and dust and providing privacy. Municipalities could benefit from the creation of hedgerows on roadsides where winter snow drift causes road closures. When a hedgerow is in place, the drifting snow collects at the base of the plants instead of on the street.

Ideally, hedgerows are composed of an assortment of native trees and shrubs, underplanted with flowering perennials and native grasses to increase plant diversity and curb appeal. Choose a broad variety of plants for the best results (see the list on facing page). For added bonus, you can include a few edible plant species in your hedgerow to provide yourself with a tasty benefit.

A hedgerow of red twig dogwood stands out in the wintertime.

The biggest challenge in hedgerow construction is to create a planting that doesn't look messy. Though farmers and other rural residents may be able to install a casual hedgerow that is only minimally maintained, in suburban and urban areas, gardeners must be careful to plant with purpose and properly maintain their hedgerow to avoid issues with neighbors and community organizations. Slightly more formal hedgerows are called for where neighbors are close; these hedgerows may need to be regularly weeded, mulched, and otherwise maintained.

Some great plants for hedgerow construction in North America include the following:

+ American holly (*Ilex opaca*)
+ American persimmon (*Diospyros virginiana*)
+ Birch (*Betula* spp.)
+ Black-eyed Susans (*Rudbeckia* spp.)
+ Blueberries (*Vaccinium* spp.)
+ Coneflowers (*Echinacea* spp.)
+ Elderberry (*Sambucus* spp.)
+ Meadowsweet (*Spiraea* alba)
+ Ninebark (*Physocarpus* spp.)
+ Pawpaw (*Asimina triloba*)
+ Prairie dropseed (*Sporobolus heterolepis*)
+ Redbud (*Cercis canadensis*)
+ Serviceberries (*Amelanchier* spp.)
+ Sourwood (*Oxydendrum arboreum*)
+ Sumac (*Rhus* spp.)
+ Sunflowers (*Helianthus annuus*)
+ Switchgrass (*Panicum virgatum*)
+ Viburnums (*Viburnum* spp.)
+ Virginia magnolia (*Magnolia virginiana*)
+ Witch hazel (*Hamamelis* spp.)
+ Yarrow (*Achillea* spp.)

Yarrow (*Achillea* spp.)

Elderberry (*Sambucus* spp.)

Witch hazel (*Hamamelis* spp.)

A carpet of low-growing plants, such as this sweet alyssum surrounding carrot plants, can help deter pests that lay eggs in the soil, including the carrot root maggot fly.

Hedgerows are linear and can be any length and width, though a minimum width of 10 to 12 feet is best. Larger areas offer the opportunity for greater plant diversity and an increase in the benefits a hedgerow can provide.

Low-Growing Cover Crops for Soil-Dwelling Pests

The second way companion planting can physically impede pest movement is to use low-growing plants to block access to the soil for those pests that lay eggs, pupate, or dwell in the ground.

Many different pest insects spend part of their lifecycles in the soil as eggs, larvae, pupae, or adults. Flea beetles and cucumber beetles, for example, feed on plants aboveground as adults, but their larvae live underground and consume the roots of host plants. Numerous species of root maggot flies, including carrot, onion, and cabbage maggot flies, and several species of aphids eat plant roots, too, as do a handful of other pests. Cabbageworms and cabbage loopers, tomato and tobacco hornworms, yellow-striped armyworms, tomato fruitworms, corn earworms, and various other species of caterpillars drop off their host plant when it's time to pupate, then burrow into the ground and form their cocoons within the soil.

It's been shown that the presence of low-growing cover crops and living mulches decrease the number of pests that spend part of their lifecycle underground. Female pests who lay their eggs in the soil, such as cucumber beetles, flea beetles, and root maggot flies, have a harder time reaching it. And by blocking access to earth, the ability of other pests to enter subterranean pupation is greatly reduced. Whether it's an "official" living mulch planting such as crimson clover, oats, white clover, or cowpeas or a low-growing plant such as thyme or sweet alyssum, simply having a living plant covering the soil limits the number of these pests found around susceptible plants.

JUST MIX IT UP

The final way companion planting can help gardeners manage pests is one that I preach throughout the pages of every chapter of this book: diversity, diversity, diversity. Yes, employing the specific plant partnerships found within this chapter will help you target specific pests. But the truth is, you'll encourage a balanced garden with an overall reduction in pests and an improvement in the number and diversity of pest-eating beneficial insects simply by ensuring that your garden is as diverse as possible.

DISEASE MANAGEMENT

Suppressing Disease through Plant Partnerships and Interplanting

It's estimated that in the northeastern United States, where I reside, 10 to 15 percent of the vegetables grown for fresh consumption and processing are lost to soilborne diseases. That's a devastating amount of loss to the agriculture industry. For home gardeners, losses due to soilborne pathogens are equally as destructive. If you've ever faced the following common plant diseases, you know how destructive they can be:

+ Stem or crown rots, including tomato blights, cottony rots, southern blight, damping-off, stem cankers, and fruit rot, among many others
+ Root rots
+ Wilt disease
+ Bacterial diseases such as potato scab, fire blight, cucumber wilt, soft rot, and others

In this chapter, we'll look at companion planting techniques that limit the occurrence of these and other common plant diseases.

COVER CROPS *and* LIVING MULCHES

The integration of cover crops and living mulches into the farm field or home garden is one of two companion planting techniques that suppress certain plant diseases. Studies have shown reduced disease rates when the cover crop is planted the season prior to the vegetables and then tilled in. But perhaps more important to home gardeners, cover crops have also been found to reduce disease prevalence when their residue is left in place, as well as when they're grown as a living mulch.

Exactly how disease suppression occurs when using cover crops and living mulches is a topic of much research. It's believed that soilborne plant diseases may be suppressed by the increase in the diversity and activity of soil microorganisms that cover crops and living mulches (and other sources of organic matter) introduce to the soil. In fields and gardens with a rich array of soil-dwelling microbes, pathogenic microorganisms have more competition for nutrients. And beneficial microorganisms may also produce compounds that affect the ability of pathogens to thrive.

There are six main ways the use of cover crops and living mulches can help manage soil diseases, each of which we'll explore below.

Generating Antifungal Compounds or Conditions in the Soil

Fungistasis is the term for the inhibition of fungal organisms. It's the result of either the presence of certain antifungal compounds within the soil or the decrease in certain nutrients necessary to fuel the growth of the fungal pathogen. Companion planting is an important and emerging approach to controlling fungal plant pathogens. The results of several case studies show that the ability of a soil to suppress fungal pathogens declined with the decrease of soil bacterial diversity, which suggests that greater soil bacteria diversity can lead to fewer plant diseases. The extent of this disease suppression, however, is highly variable and depends on many factors, such as the soil type, the type of cover crop being used, and the climate. As research and experimentation in this area reveal more information, our understanding of the mechanisms behind disease suppression continues to expand.

Certainly not all cover crops and living mulches are capable of increasing fungistasis in the soil, but several have been well studied and have been found to influence the prevalence of disease. Mustard greens (*Brassica juncea*), for example, contain compounds called glucosinolates that have fungicidal properties. In addition, after decomposition, the compounds released from the mustard greens increase the activity and diversity of soil microorganisms, leading to a further reduction in disease pathogens. Growing mustard greens as a cover crop prior to growing potatoes has been shown to reduce the rate of rhizoctonia (root rot) infection. Cover crops of Sudan grass and mustards have also been shown to have "biofumigant" properties when tilled into the soil, resulting in a reduction of verticillium wilt on tomatoes.

Supporting Beneficial Microorganisms

Another way cover crops and living mulches can impact soil-dwelling diseases is through their ability to support the beneficial microorganisms that are antagonistic to pathogenic species. This can occur as general disease suppression, where competition for resources among the microorganisms decreases the ability of pathogenic organisms to infect a plant, or it can occur as specific disease suppression, which results from an increase in the particular soil microbes that act directly against pathogenic microorganisms. For example, several beneficial soil-dwelling fungi species are known to combat pythium, fusarium, and other common vegetable pathogens.

Regardless of whether they serve as general or specific suppressors, the diverse array of microorganisms within the soil affect the presence of soilborne diseases. It's widely acknowledged in the scientific community that microorganisms in the root zone can increase a plant's ability to tolerate stress, fuel plant growth, and suppress the development of disease. There's also good evidence that the microorganisms associated with roots can alter the plant's natural defense system, encouraging disease resistance.

Since cover crops and living mulches influence both the amount and species of microorganisms living in the soil, their presence limits disease by improving the overall health of the plant by increasing its ability to absorb nutrients and water.

A cover crop of sudangrass

The developing fruits of vine crops, when left in direct contact with the soil, have an increased risk of soilborne disease exposure.

Reducing the Splash-Up Effect

A third way cover crops and living mulches impact disease rates is by creating a physical barrier that reduces soil splash. Many soilborne diseases make their way onto the foliage or fruits of vegetable plants through the splash-up effect. When rain or irrigation water hits the soil, it bounces up onto the lower leaves of a plant, possibly bringing disease-causing fungal spores with it. One study showed that the foliar disease septoria leaf spot was greatly reduced on tomatoes planted into the residues of a cover crop. The same was shown for fungal rots on pumpkins when they were planted in the residue of a rye grass cover crop. For fruiting plants such as pumpkins, squash, watermelons, and cucumbers, whose fruits sit on the ground, cover crop residues or living mulches reduce the rates of soil splash, and subsequently the rates of disease infection, by forming a protective layer between the developing fruit and the soil. It also keeps the developing fruit from making direct contact with the soil.

Limiting Certain Insect-Transmitted Diseases

Companion planting techniques that utilize cover crops and living mulches may reduce the rates of certain insect-transmitted pathogens by making it more difficult for the disease-transmitting insect to find its host plant. As discussed in the previous chapter, interplanting gardens with multiple plant species rather than having a monoculture makes it more difficult for many herbivorous insects to find their preferred host plants. Therefore, the rates of pathogen transmission are often lower in gardens with a mixed matrix of plantings, including cover crops and living mulches.

Some plant partnerships established through cover cropping and living mulches reduce insect-transmitted pathogens by increasing the numbers and diversity of beneficial insect species that help manage the disease-transmitting pest insects. We'll explore this in greater depth in chapter 7, which discusses how to use companion planting for improved biological control

Improving the Overall Health of the Soil and the Plants

In chapter 2, we looked at how companion plants are used for soil preparation and conditioning. Improving the soil fertility and structure in turn leads to healthier plants better able to resist soilborne diseases and access the nutrients and water they need to fuel their growth.

Using cover crops and living mulches reduces soil compaction, too. Heavily compacted soils promote certain root disorders caused by various environmental conditions rather than by the presence of living organisms like fungal spores or bacteria. Roots sitting in waterlogged soils, for example, can develop disorders due to their lack of oxygen. The organic matter and microorganisms present in soils where cover crops are used reduce soil compaction and therefore the occurrence of these types of disorders.

Preventing a Buildup of Pathogens in the Soil

One more surprising way a cover crop can help limit plant diseases is by restricting a soil-dwelling pathogen's ability to populate the soil. When alternated with vegetable crops, cover crops increase the amount of time between disease-prone plantings,

Companion planting can be done in containers, too. Here, a small-leaved variety of white clover is combined with squash plants.

Be Patient, Be Vigilant, Be Flexible

Sometimes cover crops can actually *increase* certain diseases. If you till in a cover crop, be sure to allow three to four weeks to pass before planting vegetable transplants in that area. Some pathogens, such as pythium, are stimulated by the addition of cover crops into the soil, so the till-and-wait period is essential. Other cover crops themselves can host pathogens that may go on to infect the vegetable crops planted into their debris. White clover and buckwheat, for example, could increase rates of bean root rot if the beans are planted too soon after the cover crop is tilled under.

Brassicas are among the best cover crops for suppressing diseases, but they can sometimes affect the mycorrhizal activity and diversity on the roots of the vegetable crops planted after them. A temporary decrease in mycorrhizal activity could, in turn, negatively affect disease suppression by decreasing the amounts of nutrients and water available to the vegetables via the mycorrhizal network.

As with most things in life and the garden, figuring things out demands experimentation and flexibility. Try various companion planting methods to see what controls diseases best in your specific growing conditions. Remember that companion planting is an emerging science with much that remains to be learned and trialed.

and since most soilborne diseases can't survive more than a few years without a suitable host, cover crops combined with crop rotation keep pathogen populations from building up within the garden. Though the length of time any given pathogen can survive in the soil varies, repeatedly planting the same disease-prone crop, or group of crops, in the same area is a recipe for repeated infection.

No matter which of the above mechanisms of disease control are present, it's important to remember that using cover crops and living mulches with the aim of limiting soilborne disease produces variable results, especially when used short term. There are many factors in play, including weather, management of the cover crop, soil conditions, and more. Long-term studies are promising, but for home gardeners in particular, 100 percent disease control will not likely occur. The goal instead is to use companion planting to naturally reduce the frequency and spread of plant pathogens to whatever extent possible.

PLANT PARTNERS *to* MANAGE SOILBORNE DISEASES

Here are some well-studied plant combinations involving a cover crop and a productive vegetable crop, all aimed at promoting natural disease suppression in the vegetable crop.

Potatoes + Oats or Winter Rye for Verticillium Wilt

Verticillium wilt is a fungal disease found in soils across much of North America. It affects the water-conducting tissue of its host, causing the plant to yellow and wilt. Premature death is often the result. When the disease occurs on potato plants, tuber size and yields are reduced. Many different plants are susceptible to verticillium wilts, and the organisms that cause verticillium wilt can live not only in the soil but also on tools and equipment or on alternate host plants.

This plant partnership is for gardeners whose potato crop has been affected by verticillium wilt in the past. A cover crop of oats (*Avena sativa*) should be grown in the area the fall before planting potatoes. If you recall from chapter 2, oats are a cool-season cover crop that's winter-killed in cold climates. Sow oats in the garden in late summer, wherever you plan to grow potatoes the following year. Come spring, leave the oat residue in place and plant the potatoes right through it. The presence of the oats causes greater colonization of mycorrhizae in the potatoes, which has been shown to displace the verticillium pathogen.

Potato plants grow through a mulch of oat residue.

Winter rye (*Secale cereale*) is an alternative cover crop used to reduce rates of verticillium wilt on potatoes. If you choose winter rye rather than oats, you can either till the rye residue into the soil three weeks prior to planting the potatoes or cut the rye down just as it comes into flower, leave the residue on top of the soil, and plant the potatoes through it.

Potatoes + Brassicas for Potato Scab

Potato scab is a disease that's found throughout the world. Though it doesn't typically kill potato plants, it does cause lesions on the skin of the potatoes and limits their storage life. Potato scab builds up in the soil with each successive potato crop, so if you plant potatoes in the same area every year, scab can become very problematic. Using a cover crop from the brassica family inhibits the pathogen that causes potato scab. Brassicas release glucosinolates, which are a specific disease suppressor for streptomyces, the organism that causes common potato scab. Mustards (*Brassica* spp.) were shown to be the best brassica for the job, while canola was the least successful brassica. For this disease-suppressing plant partnership to work, the brassica cover-crop debris must be incorporated into the soil three to four weeks prior to planting potatoes.

Verticillium wilt is a devastating disease of potatoes and other common garden plants.

Cauliflower and Lettuce + Brassicas for Verticillium Wilt and Sclerotina Stem Rot

In addition to managing potato scab, brassica cover crops have also been shown to reduce the rates of verticillium wilt in cauliflower and stem rot in lettuce. Again, for these partnerships to work, the brassica cover-crop debris must be tilled into the soil three to four weeks prior to planting the cauliflower or lettuce transplants into the garden.

Verticillium wilt in cauliflower exhibits as plant yellowing and wilting, possibly leading to plant death. Because *Verticillium* fungus is a soil-dwelling pathogen, it can be difficult to manage. Stem rot in lettuce causes the plant to develop brown, mushy decay on the stem or crown followed by white, fuzzy growth. You'll also find hard, round, black nodules on infected plants (called sclerotia), which can survive in the soil for many years. Sometimes also called lettuce drop, this disease readily spreads in wet, moist conditions. Other hosts of this disease-causing fungus in the vegetable garden include beans, celery, tomato, cauliflower, and more.

A cover crop of mustard can be an effective check on potato scab. ▶

Tomatoes + Hairy Vetch for Managing Foliar Diseases

Tomatoes are susceptible to several soilborne fungal pathogens, including early blight and septoria leaf spot, which cause lesions, spotting, and yellowing of the plant's foliage and often reduced yields. Tomatoes grown after a cover crop of hairy vetch (*Vicia villosa*) had less foliar disease than those grown in a plastic sheet mulch. They had higher yields as well. Keep in mind that hairy vetch is not the same plant as crown vetch, an aggressive invasive plant often used in roadside plantings to control erosion. Hairy vetch is far easier to manage than crown vetch, which will quickly take over a garden.

Hairy vetch is a cold-tolerant legume often used as a cover crop and is winter hardy in colder regions. It fixes large amounts of nitrogen. Planted in fall, hairy vetch can grow up to 4 feet tall the following spring. For this companion planting technique, do not till the hairy vetch into the soil. Instead, cut the plants down with a string trimmer, lawnmower, sickle, or scythe, or flatten them at ground level with a heavy lawn roller. Do this right when the first pods appear on the vetch plants in late spring or early summer. Don't wait until the pods swell, or the vetch could reseed and become problematic; but don't do it too early either, or it may not kill the vetch plants. Leave the residue in place and plant the tomato transplants through it soon after cutting.

The vetch residue forms a thick mat that's not only great for suppressing foliar diseases but also excellent at deterring weeds. A cover crop of hairy vetch has been shown to increase disease resistance in other vegetables, too.

Sclerotina stem rot in lettuce plants is reduced by using certain brassicas as cover crops.

Watermelons + Hairy Vetch for Reducing Fusarium Wilt

Fusarium wilt is a disease that restricts the movement of water throughout an affected plant. It causes individual branches or vines to turn yellow and wilt with no recovery. Sometimes only a single branch or one side of the plant is affected. Eventually, the entire plant will die. The fungi that cause the disease overwinter in the soil easily, as well as on seeds. Fusarium wilt sometimes affects just a few scattered plants rather than all of them.

Growing a cover crop of hairy vetch prior to planting watermelons (*Citrullus lanatus*) has been shown to reduce the rates of fusarium wilt. In one study, it increased the sugar content of the watermelons, too. This partnership seems to work because the cover-crop residues reduce the rates of stem colonization by *Fusarium* fungi.

Plant the vetch in autumn. Come spring, cut the tall, vining vetch plants down just as the first pods appear. Do not allow the seeds to fully develop. Leave the vetch residue in place, and plant your watermelon seeds or transplants right through it. The thick mat of vetch stems serves to limit pathogens, prevent weeds, and keep the ripening watermelons off the soil.

Another study showed that using a cover crop of crimson clover improved the mycorrhizal colonization levels of watermelon plants and led to a reduction in fusarium wilt.

Hairy vetch is a great cover crop for tomatoes and watermelons and also has beautiful purple flowers.

This bean plant is suffering from root rot.

These seedlings are about to be planted through a mulch of winter rye residue.

Beans + Winter Wheat or Winter Rye to Manage Root Rot

Bean root rot is a plant disease caused by a fungus (*Rhizoctonia*) that lives below the surface of the soil, making it difficult to diagnose the problem before the crop is destroyed. It affects not only beans but several other vegetables and ornamentals, too. The disease results in a reduction in plant vigor, stifled yields, and plant death. If you suspect bean root rot is affecting your garden, pull up a plant and look for small, slimy, water-soaked cankers on the roots. The size of the root system will be reduced as well. You may also find the roots to be dark in color. These fungi can live in the soil and on infected plant debris for many years, making crop rotation essential. They more commonly strike dry bean varieties than snap beans. Wet weather and compacted soils are the perfect recipe for root rot infection.

Studies have shown that growing beans in conjunction with a cover crop of winter wheat (*Triticum aestivum*) or winter rye helps reduce the rates of root rot in beans. For either winter wheat or winter rye, sow the cover crop seeds in late summer or early fall. Both species will survive the winter. Come spring, both winter wheat and winter rye should be mown down to the ground just as the plants come into flower. Leave the residue in place and then plant bean seeds through it. Though winter rye is allelopathic, this trait should not affect the germination of the bean seeds. It inhibits the germination of smaller seeds, not larger-seeded vegetables such as beans.

IMPROVING AIR CIRCULATION

There's another type of companion planting that limits the incidence and spread of the many fungal diseases that colonize the leaves and stems of a plant. Examples of these types of diseases caused by fungi are rusts, powdery mildews, downy mildews, leaf blights, and leaf spots.

Many fungal pathogens that infect plants throughout the growing season are spread via spores that move from plant to plant on air currents, animal fur, and even on human skin and clothing. The germination, establishment, and growth of these spores on the foliage of a new host plant often rely on the presence of moisture. Practices such watering in the morning so the foliage quickly dries, focusing irrigation water on the root system rather than on the leaves, and spacing plants properly to maximize air circulation are all simple ways to limit fungal diseases. With some basic but well-planned companion planting, however, these practices will be even more effective.

Companion planting can ensure there is ample air movement around each plant so the foliage can dry off quickly after rain or irrigation, or during periods of high humidity. This improvement in circulation isn't achieved because plants are spaced far apart with bare soil in between. Instead, it's because the plants are planted in layers. Tall plants, such as tomatoes, peppers, and okra, are surrounded by or interplanted with lower-growing varieties, such as carrots, beets, and bush beans. This allows for maximum use of space while still providing good airflow around each layer of plants. You can use this technique both in a traditional row-planting layout or in a *potager*-style mixed planting or raised bed.

By planting in layers, you'll maximize air circulation and cut down on diseases.

For row plantings, rather than planting three rows of tomatoes next to each other, alternate the tomato rows with a row of a shorter crop such as bush beans, or alternate the plants within the same row. In a mixed vegetable garden that uses blocks or raised beds, surround taller crops with a skirt of shorter veggies, mixing and matching species by their stature.

The improved air circulation provided by this companion planting technique not only suppresses certain foliar fungal diseases but also supports a high population of diverse mycorrhizal fungi within the soil, minimizes weed seed germination by shading the soil, and limits damage done by leaf-munching insects by reducing the ability of certain pests to find a suitable host plant among all that different foliage.

In addition, mixed companion planting like this can reduce the splash-up effect by covering exposed soil with the foliage of another plant less susceptible to soil-dwelling pathogens.

Alternate shorter growing crops, such as bush beans, with taller crops, like tomatoes, to form a layered effect and maximize air circulation around plants.

The Influence of Weather and Variety

Because fungal pathogens are highly influenced by the amount of moisture in the environment, even the best-managed vegetable gardens will face years with high fungal disease pressure due to wet weather conditions or constant high humidity. During years with frequent spring and summer rainfall, or in climates with naturally high levels of humidity, these companion planting methods may prove less effective. Or they'll need to be combined with other disease-control measures, such as the use of organic fungicides, proper garden sanitation, and conscientious maintenance practices.

It's always a smart idea, regardless of weather predictions, to plant vegetable varieties that have a noted resistance to plant pathogens. This resistance may either be bred into them or occur naturally. When purchasing transplants or seeds, pay attention to any notes on disease resistance mentioned in the seed catalog, on the seed packet, or on the pot tag. Disease resistance is indicated by code letters somewhere before or after the varietal description. For example, "CMV" means that variety is resistant to cucumber mosaic virus, while "F" denotes resistance to fusarium wilt. You can find the key to the code in the seed catalog or on various websites, including that of the International Seed Federation. Many organizations are trying to make this pathogen resistance code universal, but some seed catalogs have their own codes.

BIOLOGICAL CONTROL

Plant Partners That Attract and Support Pest-Eating Beneficial Insects

Though we gardeners tend to focus our insect-related efforts on controlling pest species, the truth is that less than 1 percent of the millions of identified insect species on the planet are classified as agricultural or human pests. The vast majority of the bugs we come across in our yards and gardens aren't harmful to us or to our landscape. Instead of munching on our plants (or pets or kids), they're serving another function in the ecosystem. Perhaps they're decomposers or pollinators — or maybe, if you're lucky, they're predators or parasitoids who are consuming and controlling the pest species nibbling on your cherished plants. It's a bug-eat-bug world out there, after all.

Instead of focusing so much time, effort, and money on battling the pest species, it's time to switch our focus to encouraging the beneficial insects that naturally keep pest populations in check. Doing so brings a natural balance back to the garden and results in a reduced need for pesticides and other pest-management strategies.

WHAT IS BIOLOGICAL CONTROL?

The science of biological control, or biocontrol, uses one living organism to help control the population of another. In essence, it means encouraging the "good" bugs to help manage the "bad" ones. Before we examine how companion planting can enhance biological control, it's important to understand the three main types of biocontrol and how they work.

The first two types of biological control involve intentional insect releases. On a small scale, it's known as augmentative biocontrol. You've probably seen containers of ladybugs on the shelf at your local garden center or spotted a sign at your local botanical garden about how they've released a special parasitic wasp to control a particular pest. While many of these augmentative releases are effective, especially in the contained environment of a greenhouse or conservatory, others prove far less helpful. Augmentative biocontrol is a highly involved science, especially since the majority of insect predators feed on only a specific pest or group of pests. You have to know that what you're releasing actually eats your problematic pest. Otherwise, it's wasted time and money.

Biocontrol experts know precisely which predators help manage which pests; most of these insects are useful primarily in commercial growing operations. As for the container of ladybugs on the garden center shelf, don't buy them. In most cases, they are convergent ladybugs (*Hippodamia convergens*) that are wild collected from their overwintering sites on sunny mountaintops in the western United States and shipped around the country for sale. The practice disturbs wild populations and potentially spreads disease to indigenous ladybug species in your garden. If you're going to practice augmentative biocontrol at your home, purchase beneficial insects only from commercial insectaries where they're reared in captivity. Staff at the insectary can help you choose the correct beneficial insect to best control your exact pest issue.

Lacewings from an insectary are ready to be released into a garden.

On a large, outdoor scale, intentional insect releases fall under the umbrella of classic biological control. You may have occasionally heard on the news about a governmental agency releasing some kind of specific predatory or parasitic insect into the environment to help control an invasive insect that was introduced to our continent where it has no natural predators. Classic biocontrol has proven to be a very useful tactic in many situations, though there have been a few missteps along the way. For example, the Asian multicolored ladybug (*Harmonia axyridis*) was imported to North America to help control tree pests. It has since become a nuisance pest itself when it enters homes and other structures to overwinter. When Asian multicolored ladybugs were originally released, there were few steps in place to carefully study in advance the predator and its effects on the local ecosystem. Now scientists spend years studying a particular insect's aptitude for biocontrol and any possible effects on nontarget species before they even consider releasing it into the environment.

Though the Asian multicolored ladybug is an example of biocontrol gone wrong, there have been far more positive outcomes than negative from large-scale classic biological control techniques. For example, the Mexican bean beetle is well managed by a little introduced wasp called the pedio wasp (*Pediobius foveolatus*), cottony-cushion scale was eliminated from California citrus farms by the release of the vedalia ladybug (*Rodolia cardinalis*), control of the elm leaf beetle was the result of two different parasitic eulophid wasps, and the management of euonymus scale is the result of the release of a non-native, host-specific ladybug in several regions of the United States.

The final type of biological control is the one that is most important to the objectives of this book. It's called conservation biological control. Conservation biocontrol refers to protecting and promoting the beneficial insects already in your landscape by reducing or eliminating pesticides, by creating a habitat that's favorable to the insects, and by providing them with any necessary food sources, *not* by intentionally releasing insects into the environment. Modifying your backyard habitat to increase numbers of beneficial insects is an excellent way to help manage certain pests — and companion planting can be a valuable tool for doing it.

A small parasitic wasp feeds on dill flowers.

ENHANCING NATURAL PEST PREDATION

Predatory insects capture and consume other insects directly, while those that are *parasitoidal* (often incorrectly referred to as parasitic) use other insects to house and feed their developing young, eventually killing the host insect in the process (true parasites leave their host alive). Both of these groups of insects consist of tens of thousands, if not hundreds of thousands, of different insect species. Commonly called "beneficial insects" or "natural enemies," these insect predators and parasitoids are to be encouraged in the garden as a natural means of pest control.

Many beneficial insects, however, require more than the protein found in their prey to survive. Ladybugs and lacewings, for example, often can't reproduce without the carbohydrates found in nectar, and while larval parasitic wasps feed on insects like aphids and tomato hornworms, the adult wasps rely on the sugars found in flower nectar. Then there's the adult hover fly, whose young feast on aphids while they consume protein-rich pollen and carb-infused nectar to survive. There are certainly predatory insects who eat nothing beyond their insect prey (robber flies, dragonflies, and ground beetles, for example), but the majority of beneficial insects require pollen and nectar from plants at some point in their lifecycle. This is where companion planting comes in.

There are several ways companion planting can be used to enhance pest control through the encouragement of beneficial insects:

- By partnering pest-prone plants with plant species known to provide nectar and pollen to the species of beneficial insects most likely to prey upon the pests
- By ensuring there are as many beneficial insect–friendly plants in the garden as possible to keep the population of beneficial insects well fed so they stick around to control future pest outbreaks
- By including companion plants that create year-round habitat for beneficial insects

The actionable companion planting strategies that follow serve to either provide habitat for predators and parasitoids or to provide these insects with the pollen and nectar forage they need to thrive — both of which, in turn, encourage these good bugs to stick around and help manage pests.

A beneficial syrphid fly drinks nectar from fennel flowers.

"Banker" Plants: An All-You-Can-Eat Bug Buffet for Beneficial Insects

One final way companion planting encourages beneficials is through the use of "banker" plants. Banker plants are essentially sacrificial companion plants that serve as hosts for pests, providing beneficial insects with a continuous supply of prey when vegetable plants meant for harvest are pest-free. Banker plants ensure the population of beneficial insects is able to reproduce and stay present in the garden even when the pests aren't problematic to crops. For example, grain crops are sometimes used as a food source for aphids to keep populations of parasitic wasps, lacewings, ladybugs, and other aphid predators high so that when vegetable or fruit crops start to host an aphid population of their own, the natural enemies are already there. The use of banker plants is currently focused on farm, orchard, vineyard, and greenhouse production, though perhaps in the future, home gardeners will find this strategy useful as well.

PLANT PARTNERS *to* ATTRACT BENEFICIAL INSECTS

Sometimes, a pest-prone vegetable plant can be paired with another plant that's known to provide food or habitat for predators of whatever particular pest is feeding on it. In most cases, beneficial insects do not have the specialized mouthparts required to access nectar from deep, tubular flowers. Instead, they have mouthparts that can drink nectar or eat pollen only from blooms with shallow, exposed nectaries, which means, of course, that not just any flowering plant will do. Many trials and studies have been conducted, and continue to be conducted, to discover which plants are best at attracting which beneficials. These studies give us excellent, actionable information we can then use to make decisions about which plant combinations will work to limit pest numbers.

This tobacco hornworm has been parasitized by a wasp known as the cotesia wasp, as evidenced by the presence of pupating wasp cocoons on the outside of its body.

In addition to employing specific plant partnerships in your garden, it's critical that you take a more general approach to supporting as many natural enemies as you can. Because there's such a broad diversity of pests that commonly feed on vegetable and fruit plants, protecting your garden from damage involves incorporating plants that will provide for beneficials everywhere across your landscape, not just in the immediate proximity of a pest-plagued plant. The greater diversity of plants you have in your garden, the more beneficial insects your garden will be able to support. Diverse landscapes have been shown time after time to be more stable environments with less pest pressure and fewer outbreaks. To this end, you should aim to include plenty of flowering herbs, annuals, and perennials right in your vegetable patch to create a mixed habitat. The companion planting methods detailed in this chapter focus primarily on specific partnerships, but my previous book, *Attracting Beneficial Bugs to Your Garden: A Natural Approach to Pest Control*, offers detailed insight into how to implement a more general approach.

The following companion planting ideas are designed to foster biological control by providing various species of beneficial insects with pollen and nectar. Since most pest-eating beneficials require plant-based food at some point in their lifecycle, it's essential that the right plants be present in your garden — ones they can easily feed from. Nonetheless, the suggestions here are still fairly broad based, especially when compared to partnerships presented in previous chapters. This is because the plants that are particularly adept at providing food for beneficials can do so for many different species, not just one. So instead of outlining how one plant partners well with another, here I've partnered a specific plant that's good at feeding beneficials with a pest insect that's being targeted. Read on and you'll see what I mean.

Lettuce and Other Greens + Dill and Fennel for Aphid Control

There are hundreds of different species of aphids found in North America. Some aphids are host specific, meaning they can feed only on one plant species or family of plants. Other aphid species are more general in their feeding habits. For example, the aphids that feed on peach trees are not the same species that attacks your milkweed plants or sucks the sap from your spruce trees. However, all species of aphids are common prey for an incredible diversity of beneficial insects, including tiny parasitic wasps, ladybugs, lacewings, spiders, soldier beetles, syrphid flies, minute pirate bugs, damsel bugs, and many others.

No matter what species of aphids you're dealing with and which plant they're attacking, you can draw many of their predators by planting members of the carrot family (Apiaceae), including dill, cilantro, angelica, golden Alexander, and fennel. The blooms of these plants consist of thousands of tiny flowers gathered into an umbrella-shaped inflorescence. Each of these tiny blooms has a shallow, exposed nectary — the perfect fit for almost all of the aphid-eating beneficial insects mentioned above. Scatter members of the carrot family all throughout your vegetable garden for improved and natural aphid management. As you learned in chapter 5, as an added bonus you'll help "hide" host plants from flying pests by interplanting crops with these fragrant herbs.

Interplanting leafy green crops, such as kale, with flowering members of the carrot family, supports aphid-eating beneficial insects and limits aphid damage.

Plant fennel and other members of the carrot family near lettuce to lure beneficial insects and increase aphid predation.

Because members of the carrot family attract a variety of beneficial insects, they can be used with a number of other plants to help control specific pests. Here are a few more companion planting strategies that fit under this umbrella:

+ Combine **cilantro with cabbages** for aphid management.
+ Interplant **dill with cole crops** to control cabbageworms.
+ Interplant **dill or cilantro with eggplants** to control Colorado potato beetle (see next section for more detail).

Eggplants + Dill or Cilantro to Control Colorado Potato Beetle

A study examining the presence of Colorado potato beetles in eggplant fields found that when the eggplant rows were alternated with rows of flowering dill (*Anethum graveolens*) and cilantro (*Coriandrum sativum*), there was a significant increase in the number insects that prey upon Colorado potato beetles. As a result, the interplanted fields had less pest damage from these insects than a control plot without the flowers.

In the study, the dill and cilantro were not planted immediately adjacent to the eggplant but rather in alternating strips. In a home garden, where you may be growing just a few eggplants as opposed to long rows of them, a nearby planting of dill and cilantro may be all you need to lure in enough predatory insects to keep Colorado potato beetles and their larvae in check.

Flowering dill interplanted with eggplants leads to a reduction in Colorado potato beetles due to the increase in predatory insects drawn to the nectar and pollen of the dill blooms.

Members of the aster family, such as this cosmos, make great plant partners with cole crops. They serve as nectar sources for many species of beneficial insects that feed on aphids.

Cole Crops + Black-Eyed Susans and Cosmos **for Aphid Control**

Though black-eyed Susans (*Rudbeckia* spp.) and cosmos (*Cosmos bipinnatus*) were the species used in the study I based this companion planting strategy on, it's likely that similar results can be achieved by extending this technique to all members of the aster family (Asteraceae). Combining plants of the aster family (which includes beauties such as coreopsis, yarrow, sunflowers, Shasta daisies, zinnias, and asters, to name just a few) with cole crops was found to reduce the number of aphids present. This is due to the prowess of asters at supporting various species of beneficial insects that dine on aphids. A defining feature of the aster family is a central disk composed of many small flowers (called disk flowers) surrounded by a ring of ray flowers we call petals. Though a sunflower may look like one big blossom, it's actually many thousands of tiny flowers put together to form the central disk. As you may have surmised by now, each of those tiny flowers provides nectar and pollen and is the ideal size for many different species of pollinators and other beneficial insects.

By planting members of this plant family in vegetable gardens — in particular, around cole crops like cabbage, broccoli, cauliflower, and kale — biological control is enhanced and aphid numbers are kept in check.

Various Vegetables + Carrot- and Mint-Family Herbs **to Control Caterpillar Pests**

Along the same lines as the previous two companion planting strategies, this one utilizes a group of plants to lure in a few different types of parasitic wasps that use various vegetable pests as hosts for their developing young. Flowering herbs in the carrot family, including dill, cilantro, caraway, anise, and fennel, as well as the mint family (Lamiaceae), including sage, marjoram, oregano, lemon balm, rosemary, and thyme, are adept at supporting these parasitic wasps by providing them with nectar.

If you want to control caterpillar pests such as tomato hornworms, tomato fruitworms, and diamondback moth caterpillars, plant a collection of these herbs in close proximity to your vegetable plants. Most parasitic wasps are host specific, meaning they only use one individual species or genus of insect to house and feed their developing young. But no matter what insects they use as hosts, all adult parasitic wasps rely on nectar for their energy needs. When the time is right, female parasitic wasps insert multiple eggs just under the skin of their host insect. The resulting larvae then feed on the host insect, which dies soon after the larval wasps pupate into adults.

The small flowers of cilantro plants provide nectar and pollen to the parasitic wasps that help control numerous pest caterpillar species in the garden.

Sweet alyssum makes an excellent companion plant for lettuce. It's a nectar source for several parasitic wasp species that use aphids to house and feed their developing young. Syrphid flies also nectar on the plants.

Lettuce or Grapes + Sweet Alyssum for Aphid Control

Several studies have examined the usefulness of the common flowering annual sweet alyssum (*Lobularia maritima*) in companion planting for biological control. It's been found to be an exemplary food source for both the syrphid flies and the parasitic wasps that help manage aphids. Like parasitic wasps, syrphid flies (also called hover or flower flies) feed on pollen and nectar as adults; it's their larvae that are predaceous. Female syrphid flies lay eggs on aphid-infested plants. The tiny, maggotlike larvae that emerge then prowl around the plant consuming aphids.

The parasitic wasp species that target aphids (known collectively as aphidius wasps) act the same way as those that use caterpillars to house and feed their young, except each tiny aphidius wasp inserts just a single egg into a single aphid. The larval wasp spends its entire larval stage within the body of the aphid until it pupates. Since lettuce and grapes are both particularly prone to aphid infestations, farmers, particularly in California, have been interplanting their fields and vineyards with rows of sweet alyssum to enhance biological control of this pest. In home gardens, similar results can be achieved by including sweet alyssum along the edges of raised vegetable beds and in gardens. Underplant taller veggies with a living carpet of sweet alyssum plants for improved aphid control.

Lacy phacelia attracts predatory bugs that like to feed on garden pests.

Cole Crops + Lacy Phacelia for Controlling Cabbage Aphids and Other Pests

Lacy phacelia (*Phacelia tanacetifolia* and *P. integrifolia*) is a beautiful flowering annual native to North America that just happens to serve multiple purposes. Extremely fast growing and often used as a green manure, lacy phacelia is a great nectar source for bees and various beneficial insects. Phacelia's flowers are arranged in an unfurling spiral, positioning them perfectly for nectar access by syrphid flies, certain parasitic wasps, tachinid flies (another parasitoidal insect), and others. As a result, the increased population of predatory insects in areas where lacy phacelia is grown has been shown to result in a reduction in cabbage aphids on various cole crops.

Lacy phacelia is also an exceptional habitat plant for spined soldier bugs, minute pirate bugs, and other predatory bugs that manage various crop pests. Phacelia can be planted directly into the vegetable garden, but know that it can become problematic if the plants are allowed to drop seed. It's a fine balance between letting the plants flower long enough to provide nectar and pollen to these good bugs but not letting them go so long that they are able to disperse seeds. I suggest watching the oldest flowers on each flower stalk for petal drop. As soon as that happens, cut the plants down.

The minute pirate bug, shown here, loves to feed on thrips.

Many Crops + Crimson Clover to Control Thrips

Thrips are a problematic pest on many different fruits and vegetables, from avocados and citrus fruits to tomatoes, peppers, and raspberries. They're tiny, thin insects that feed by sucking out plant juices. Their feeding often results in a silvery netlike appearance on leaves and fruits, along with distorted growth. Often, black flecks of excrement are visible on the infested plant's foliage. Since they're so difficult to see, thrips are hard to identify and manage. Shaking a plant with a possible thrips infestation over a sheet of white paper will help you ID this pest because you'll be able to see the slender insects crawl around on the paper. Thankfully, thrips have many natural enemies, including lacewings, mites, and even other species of thrips that are predatory rather than plant eating. But one of the most common thrips predators is a tiny insect called the minute pirate bug (*Orius* spp.).

Crimson clover is ideal habitat for this little predatory insect that can make a big dent in the thrips population. Crimson clover provides these natural enemies with both nectar (its flowers are nectar packed) and habitat. In addition, crimson clover may support other thrips predators, such as lacewings and parasitic wasps. Interplant thrips-prone vegetables with crimson clover, or use it between or under crop rows as a living mulch (see page 54).

PLANT PARTNERS *for* HABITAT CREATION

While providing food for beneficial insects in the form of prey, pollen, and nectar is critical, it's also important to provide these critters with plants they can use as habitat. Whether it's to shelter them from the elements, to protect them from other predators higher up on the food chain, to provide egg-laying habitat, or to give them a place to hunker down for the winter inside a hollow stem or under plant debris, plants that create habitat for beneficial insects are just as necessary as those that provide food. Overwintering habitat is especially important, since the overwintered beneficial insects are the ones who keep early-season pest outbreaks in check in their beginning stages rather than when pest populations have already exploded. Habitat plants should be left to stand all winter long and should not be cut back until springtime, when temperatures are regularly over 50°F (10°C) and the overwintering insects have emerged and begun to prowl the garden for pests.

Broccoli + Crimson Clover to Attract Predaceous Spiders

Though they aren't technically insects, spiders are among the most adept pest-eating critters in your backyard. While you may only see a few web-spinning spiders, there are scores of seldom-seen spider species that do not spin webs. Instead, they hunt for their prey by crawling around the ground and on vegetation in search of their next meal. These hunting spiders are known as cursorial spiders, and they do the lion's share of their work at night. Head out to the garden at midnight with a flashlight in hand, and if you don't scare them away first, you'll spy spiders munching on cabbageworms, asparagus beetle larvae, Mexican bean beetle larvae, and many other common pests.

This specific, well-studied plant partnership uses crimson clover as a habitat plant for spiders (they love hiding in thick, dense plant growth) where broccoli is being grown, but you could translate the partnership to other vegetable plants as well, particularly those in the cole crop family. For the highest level of success, either alternate rows of crimson clover with broccoli or intersperse the clover around and through the broccoli plants.

Wolf spiders dine on many common pests and often take shelter in dense plant growth.

Lettuce + Bunchgrasses to Attract Ground Beetles

One extensively studied companion planting technique that's been used by some farmers in the United Kingdom and Australia for many years is the creation of beetle banks: long, raised berms of soil planted densely with native bunchgrasses. (These are grasses that grow in a tuftlike fashion, rather than via a spreading habit. Examples would be switchgrasses, carex, and bluestems.) Beetle bank rows are alternated with crop rows in farm fields; typically, one beetle bank for every three or four crop rows. Native bunchgrasses are the ideal habitat for pest-eating ground beetles, of which there are hundreds of species.

Ground beetles eat slugs, snails, pest caterpillars, various larvae, and many other pests.

Researchers at Oregon State University are experimenting with the use of beetle banks and recommend that homeowners establish one on a smaller scale near home vegetable gardens. Called a beetle bump, these raised circles can be placed in the center of a vegetable garden or somewhere close to its periphery and planted with native bunchgrasses. They're easy to build by mounding soil 18 inches high to create a circular area at least 4 feet in diameter. Then plant the area with three or four different species of native bunchgrasses, spacing them fairly close together. The only maintenance required is irrigation until the plants are established and a once-a-year mowing down to 6 inches after the grasses have gone to seed in late fall. Leave the clippings to provide winter ground beetle habitat. The ground beetles living in your beetle bump, which feast on slugs, snails, pest caterpillars, various larvae, and many other pests, take shelter in the dense stems of the bunchgrasses during the day and come out into the garden at night to seek their prey.

Beetle banks and bumps are perennial and permanent, and they have been proven to be especially useful when used in areas where lettuce is grown. Smaller ground beetle species prey on lettuce aphids, and larger species prefer the slugs and snails that enjoy succulent lettuce leaves.

A beetle bump is a garden-sized version of a beetle bank, shown here. This row of bunchgrass between crop rows provides habitat for pest-eating ground beetles.

Low-growing flowering plants, such as this oregano, make great habitat for ground-dwelling beneficial insects.

Low-Growing Plants to Attract Ground-Dwelling Beneficials

Though we're not talking about one specific plant or family of plants for this partnership, it's still an excellent one to employ in vegetable gardens. Many pest-eating beneficial insects take shelter under the skirts of low-growing plants, some during the day, others at night. Natural enemies such as cursorial spiders, big-eyed bugs, ground beetles, minute pirate bugs, and more find low-growing herbs, ground covers, and perennials to be good protective habitat. You can create this type of habitat by interspersing a combination of short and spreading annuals, perennials, ground covers, and more into or around the vegetable garden. Oregano, sweet alyssum, gem marigolds, tarragon, basket-of-gold, thyme, and other plants that flop and rest on the ground are all excellent choices.

Hollow-Stemmed Perennials for Winter Habitat

Again, while we're not talking about a 1:1 specific plant partnership here, the practice of including any number of species of hollow-stemmed perennials into the vegetable garden or overall landscape is indeed a method of companion planting aimed at biological control. Many of the pest-eating beneficial insects we want to encourage in our gardens need habitat during the winter months when they shift into a natural period of rest akin to hibernation. Hollow-stemmed perennials, such as monarda (bee balm), phlox, heliopsis, and the like, are preferred sheltering sites for these animals; when the cold temperatures of late autumn arrive, they crawl down into the stem cavity. Even if the plant stem isn't fully hollow, some beneficials (and many species of native pollinators, too) easily excavate the pith from inside the stem and crawl in to escape the elements and predators.

It's no longer recommended that perennials be cut down and raked out in fall. Instead, do your garden cleanup in spring when temperatures are regularly reaching 50° to 60°F (10° to 15°C), after the insects have emerged.

Let perennials and ornamental grasses stand through the winter. Not only are they beautiful, they also serve as overwintering sites for many species of pollinators and other beneficial insects.

Complex Relationships

Although companion planting for improved biological control has been shown to reduce pest populations in many studies, numerous different factors are in play, as usual. Individual insects may respond differently to certain plants based on their level of specialization, making this strategy both fascinating and multifaceted. The complex and diverse relationships between insects and plants mean there's still much to be learned. Factors ranging from the synchronicity of companion plant flowering with pest outbreaks, to how far certain natural enemies are willing or able to travel, to how readily predators are able to find their prey in a complex and diverse garden all play a role in the level of success of the strategies used in any given situation. Still, there's no doubt that incorporating companion plants aimed at boosting biological control offers an effective way to keep pest numbers in check.

A robber fly dines on an adult imported cabbageworm.

Hedgerows of Closely Planted, Permanent Woody Vegetation

We discussed hedgerows in chapter 5 for their ability to restrict pest movements, but installing a hedgerow is also another means of biological control. Again, hedgerows are dense plantings of multistemmed shrubs. Often herbaceous perennials are included. The value of hedgerows in creating good bug habitat has been well studied. When placed around or near vegetable gardens and fields, hedgerows provide habitat and overwintering sites for a diversity of beneficial insects. If flowering woody plants are included in the hedgerow, a source of pollen and nectar is available as well.

Research at the University of California, Davis, has examined how far beneficial insects migrate from the hedgerow and out into the surrounding landscape. It was found that many of them could travel 80 feet or more, which means that even if the garden is a good distance away from the hedgerow, the benefits will still be achieved.

Ideally, hedgerows should include as many native plant species as possible. If you can select natives that produce fruit after flowering (such as viburnums, elderberries, blueberries, hollies, and beautyberries), you'll be providing a food source for birds and other wildlife in addition to improving biological control rates.

Fruiting shrubs such as blueberries, elderberries, and hollies make exceptional hedgerow plants.

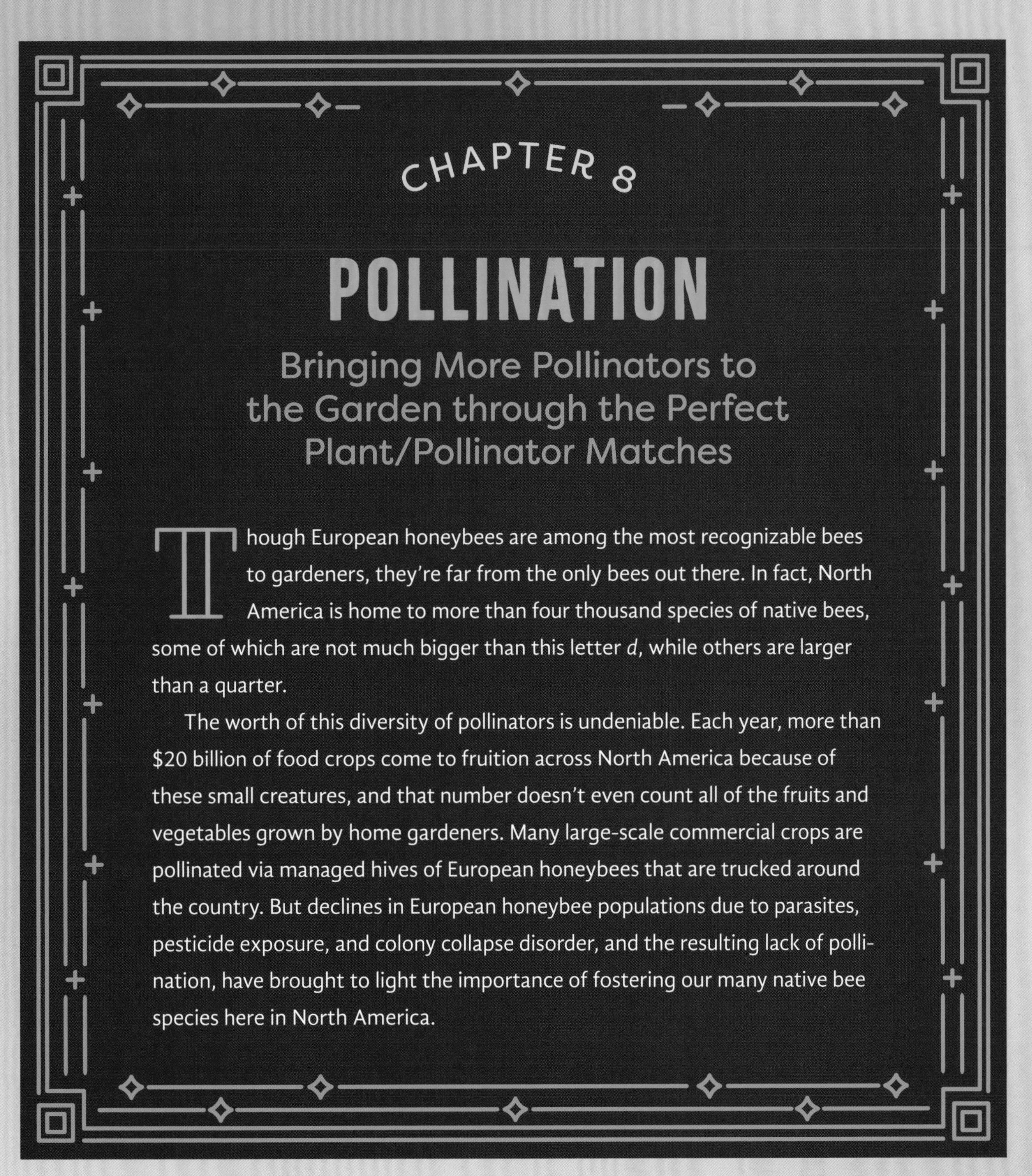

CHAPTER 8

POLLINATION

Bringing More Pollinators to the Garden through the Perfect Plant/Pollinator Matches

Though European honeybees are among the most recognizable bees to gardeners, they're far from the only bees out there. In fact, North America is home to more than four thousand species of native bees, some of which are not much bigger than this letter *d*, while others are larger than a quarter.

The worth of this diversity of pollinators is undeniable. Each year, more than $20 billion of food crops come to fruition across North America because of these small creatures, and that number doesn't even count all of the fruits and vegetables grown by home gardeners. Many large-scale commercial crops are pollinated via managed hives of European honeybees that are trucked around the country. But declines in European honeybee populations due to parasites, pesticide exposure, and colony collapse disorder, and the resulting lack of pollination, have brought to light the importance of fostering our many native bee species here in North America.

Why SUPPORT NATIVE BEES?

Though our native bees aren't suffering from colony collapse disorder, the other perils besieging the introduced European honeybee — namely, the loss of nectar forage, nesting habitat, and overwintering habitat — have taken a huge toll on many of our native bee species. Unlike European honeybees, our native bees do not spend the winter in a protected hive with other individuals, nor do they breed within a colony. Instead, females build small brood chambers in a hollow stem or small hole in the ground and lay just a few eggs each. Yes, some bee species (bumblebees, for example) do form small colonies of a few dozen individuals, but most native bee species are solitary.

Native bees are more efficient pollinators than European honeybees, and we'd be smart to encourage them in our yards and gardens. It takes 250 female orchard mason bees to pollinate an acre of apple trees, a task that requires 15,000 to 20,000 European honeybees. Unlike honeybees, most species of native bees are active in cold and wet conditions, and they have broader foraging habits. Most native bees are docile and don't sting. They're a diverse crew — with names like mining, digger, sunflower, mason, leafcutter, carpenter, and squash bees. Many are nondescript. But some are just gorgeous, with bright colors or stripes.

A native bee drinks from a blueberry flower.

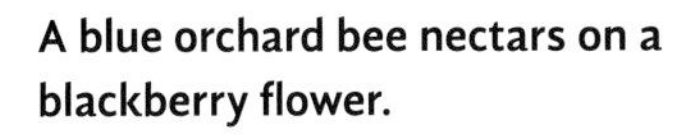

A blue orchard bee nectars on a blackberry flower.

European honeybees forage for nectar from many different species of plants.

These cucumbers have stubby ends because they weren't fully pollinated.

The decline of the overall pollinator population is taking a noticeable toll on farmers. Many cranberry, squash, and pumpkin growers; citrus groves; apple orchards; and farms of all shapes and sizes are struggling with pollination issues. Home gardeners have taken notice, reporting signs of poor pollination to their university Extension agents, online gardening forums, and Master Gardener groups. Cucumbers that are puny on one end, zucchini that never grow larger than your thumb, misshapen apples, or minimal berries on your blueberry bushes could all be signs of poor pollination.

Gardeners have a huge role to play in supporting the many species of native bees that live right outside our own front doors. Rutgers University researchers found that on larger farms, native bees are providing at least 25 percent of all pollination services. Small farmers often have more habitat for these insects, so they are probably getting even better rates of pollination from native bees. On a home garden scale, native bees perform a significant amount of pollination, especially now that European honeybees are far less prevalent in many regions. It would do all of us veggie gardeners good to encourage native bees in whatever ways we can.

The sheer diversity of our native bees is mind blowing, and their specialization is equally awe inspiring. Some native bees only pollinate one particular species or family of plants. Others are far more general in their feeding habits, feasting on nectar and pollen from a broad array of plants. Knowing which plants to include in your garden to encourage better pollination can be tough because it depends on which species of bees live where you live and what their feeding and nesting habits are.

So how can we foster and support the right kinds of bees to pollinate our home vegetable gardens? Certainly eliminating pesticides and providing these insects with nesting habitat (more on that later) play a huge role, but so does companion planting. Yep, that's right. Purposeful companion planting aimed at supporting different pollinators is a great way to enhance the pollination of many different vegetable crops.

A leafcutter bee feeds on a black-eyed Susan.

PLANT PARTNERS *to* IMPROVE POLLINATION

Here you'll find a half dozen plant/pollinator companion planting strategies to try in your home garden. While they're outlined to help a specific group of vegetable plants, for the most part, all of these partnerships benefit the garden as a whole. The greater the diversity and abundance of pollinators in your landscape, the better your chances of having exceptional pollination rates in your vegetable patch. Plus, you'll be giving some much-needed assistance to this amazing group of insects simply by providing them with some of the food and habitat that other human activities have taken away.

Let's take a look at six specific plant partnerships, including the crops to be pollinated and the best native bees for the job.

Tomatoes, Peppers, and Eggplants + Large or Hooded Flowers to Attract Bumblebees

Bumblebees (*Bombus* spp.) are an easily recognizable group of bees that visit a huge variety of crops and are active from early spring until late fall. They pollinate blueberries, squash, sunflowers, and many garden vegetables, in addition to improving pollination rates for self-fertile crops like tomatoes, peppers, and eggplants. They are social bees that live in small colonies of a few dozen individuals and need cavities to nest in. They're typically large and very fuzzy, often yellow with black stripes — though not always. About 50 species of bumblebees live in the United States. Unfortunately, some bumblebee numbers are in sharp decline due to several imported diseases that have been introduced into wild populations.

Bumblebees are valuable to crops in the nightshade family, such as tomatoes, eggplants, and peppers, because they knock the pollen loose deep within the flowers by vibrating their flight muscles very fast (it's called buzz pollination, and the buzz has the same frequency as a middle C). While tomato, pepper, and eggplant flowers are self-fertile, meaning they can pollinate themselves, the flowers need to be shaken or jostled for the pollen to be released from the anthers. The wind or even the gardener bumping into the plant's branches is often enough to cause pollen release, but the presence of

A bumblebee pollinates an eggplant flower.

bumblebees further improves pollination rates, possibly giving you better fruit set.

To boost the number of bumblebees in and around the vegetable plot, it's important to first understand how flower shape and size influence which bees visit which flower. Bumblebees are some of our biggest native bees, and they need a sturdy landing pad before they can settle on a flower. This makes plants with large, lobed, lower petals ideal. And, unlike smaller bees, bumblebees can use their body weight to pop open flowers with enclosed nectaries. In fact, bumblebees are among the only bees capable of pollinating certain hooded flowers, including snapdragons, baptisia, monkshood, lupines, and many members of the pea and bean family, including those you grow in your vegetable garden (though, like tomatoes and peppers, pea and bean flowers are self-fertile). Bumblebees have very long tongues, too, placing them on a short list of bee species capable of drinking nectar from tubular flowers like bee balm, garden phlox, and some salvias.

By planting hooded flowers and those with large "landing pads" in and around gardens where tomatoes, eggplants, and peppers grow, you'll be ensuring an ample population of bumblebees. Also, having an abundance of early-blooming plant species nearby provides food for newly emerged bumblebee queens early in spring, encouraging them to set up shop and build their nests somewhere near the garden. The offspring of those bumblebee queens will help pollinate your nightshade vegetables all summer long and well into autumn.

Bumblebees have heavy bodies, capable of popping open the closed flowers of crops in the pea and bean family.

Blueberries + Crimson Clover to Attract Bumblebees

In a Michigan study, blueberry pollination was greatly improved when the blueberry bushes were interplanted with crimson clover. Another study in Maine showed that dense plantings of pollen- and nectar-rich plants located close to blueberry bushes increased native bee populations and therefore pollination rates.

While imported European honeybees have been the traditional pollinators of farmed blueberries, native bees provided pollination services to these native shrubs long before European honeybees ever

Crimson clover and other nectar-rich plants can help bring pollinators to blueberries.

arrived on this continent. The two main native pollinators for blueberries are mining bees (*Adrena* spp.) and bumblebees, though other native bees feed from their flowers, too.

Most types of blueberries require a cross-pollination partner of another variety to produce more and larger berries. This pollen exchange is facilitated entirely by bees. Bumblebees are four times more effective at pollinating blueberries than honeybees (buzz pollination again!), and the timing of the blueberry bloom coincides with the emergence of overwintered bumblebee queens. In early spring when blueberries are in bloom, if you spy a bumblebee foraging on the flowers, it's most likely a fertilized queen who has spent the winter under a log or a pile of leaf litter somewhere nearby. As the queens forage, they also begin to build their new colonies, making blueberries a valuable early-season food source for this wonderful native bee.

Squash + More Squash to Attract Squash Bees

Squash bees (*Peponapis* spp.) are a unique group of bees consisting of about 13 North American species (the species *P. pruinosa* is found across much of the United States and as far north as southern Canada). They are all about the size of a honeybee, often with drab brown coloration and soft striping. In fact, they look surprisingly like honeybees. But unlike honeybees, during the growing season squash bees sleep inside of squash blossoms, not in a hive. If you go out to the garden in the evening after the squash blossoms have closed and find a sleeping bee when you peek inside a bloom, that's a male squash bee.

Squash bees specialize in (surprise!) pollinating members of the squash family, including pumpkins, melons, squashes, cucumbers, watermelons, zucchini, and gourds, though there are other species of bees that perform the job, too. However, a survey by the U.S. Department of Agriculture found that more than 80 percent of squash pollination on studied farms was performed by *P. pruinosa*. Squash bees are often abundant where squash is regularly grown all across the United States (except for parts of the Pacific Northwest), but they are less prevalent in home gardens.

Oddly enough, since these bees are specialists that feed only on the pollen and nectar of members of the squash family, one of the main ways to ensure you'll have a substantial population of them is to plant more members of the squash family! Since this diverse plant family includes many wonderful garden crops, planting more is great for your kitchen, too. If you're worried about running out of space, select compact or bush-type varieties of squash, cucumbers, melons, and pumpkins. They'll take up far less garden real estate. Consider growing the plants up a trellis as another space-saving option.

Also, since squash bees build solitary nests in the ground, often right underneath the squash plants themselves, tilling or turning the soil disturbs their brood chambers. Switching to no-till gardening (as described in chapter 2) preserves their nesting habitat. A study in Virginia that looked at pumpkin and winter squash pollination found that where no-till practices were in place, there were three times the number of pollinating squash bees. In fact, tilling disturbs the nests of many native bee species that create their brood chambers by excavating a tunnel or series of tunnels in bare or sparsely vegetated soil. Other ground-nesting species include bumblebees, polyester bees, mining bees, digger bees, and alkali bees. While a few native bee species are capable of stinging, most are docile and unaggressive. However, not included on this list are the many species of ground-nesting wasps, which are capable of stinging, can be quite aggressive, and do not need to be encouraged!

A squash bee gets covered with pollen inside a squash blossom.

This brilliant tiny sweat bee is covered with pollen.

Flowering Herbs + Annuals to Attract Sweat Bees

Sweat bees (*Halictus*, *Lasioglossum*, *Augochlora*, and *Agapostemon* spp.) are small pollinators ranging from ¼ to ½ inch in length. There are hundreds of species, and they can be very abundant in certain areas. Some have drab coloration; others come in brilliant colors and have a metallic sheen. All sweat bees are ground nesting, and most are solitary.

Sweat bees visit an enormous number of crops, pollinating dozens of different fruits and vegetables, from raspberries and watermelons to strawberries and fruit trees. Sweat bees are tiny but mighty, and they're important native pollinators to encourage in the vegetable garden.

In addition to pollinating edible crops, sweat bees consume nectar and pollen from many different herb and ornamental flowers, so the more flowers you have in and around the garden, the more sweat bees your garden is able to support. Allow herbs in the mint family, such as thyme, basil, rosemary, oregano, and marjoram, to flower — and plant extra so you'll still have plenty to harvest, too! Surround the edges of the vegetable garden with flowering annuals and perennials, choosing native selections when possible. Tickseed (*Coreopsis* spp.), black-eyed Susan (*Rudbeckia* spp.), mountain mint (*Pycnanthemum* spp.), ox eye (*Heliopsis* spp.), and sunflowers (*Helianthus* spp.) are favorites of sweat bees.

Members of the aster family, including this sunflower, are excellent nectar forage plants for a large diversity of native bees that also pollinate garden crops.

Greens and Root Crops + Cosmos, Sunflowers, and Daisies to Attract Small Native Bees

Many of our smaller species of native bees are responsible for pollinating various greens and root crops. These vegetables don't require pollination to produce a harvestable crop since we harvest and consume their leaves or roots, not the fruits formed from flowers. However, if you're a seed-saving gardener, ensuring there are ample pollinators in your garden translates to good rates of pollination and many viable seeds to collect and save.

Lettuce is self-pollinating and does not require an insect to move pollen from one flower to another in order to produce seeds, though the presence of pollinators typically increases pollination rates. Spinach, beets, and Swiss chard are wind-pollinated, so there's no need to worry about a lack of insect pollinators if you're saving seeds from these crops either. But many other garden greens and root crops do rely on insect pollinators to produce viable seeds, including kale, collards, mustard greens, cabbage, radishes, parsnips, rutabagas, and carrots. If you plan to save seeds from these and other insect-pollinated veggies, you'll want to have plenty of pollinators around.

Yes, honeybees can pollinate most of these crops, but many of our smaller native bees are more efficient at the job. To encourage a diversity of bees to make a home in your garden and forage for pollen and nectar on the plants from which you plan to save seeds, include many members of the aster family (Asteraceae) in your garden plans. Daisylike flowers, such as cosmos, black-eyed Susans, sunflowers, tithonia, and asters, are adept at supporting small pollinators throughout the season.

Flowering Plants to Attract Mason Bees and Mining Bees to Fruit Trees

Many long-term studies have examined the role of native bees in providing pollination services to various tree fruits and fruit-bearing shrubs. Pollinator surveys note that wild bees are often more populous in orchards than honeybees. One such study at Cornell University in New York collected close to a hundred different bee species on apple blooms. Many ongoing studies are examining what role the other vegetation in and around an orchard has on pollinator numbers and diversity.

In addition, researchers have looked at how the diversity and abundance of native bees influences fruit production. They've found that early-emerging bee species are of particular value to orchardists because fruit trees are among the earliest bloomers. The best way to encourage tree-pollinating native bees is to have a wide selection of food sources available from spring through fall. This ensures there are enough bees nesting and overwintering on-site. When spring arrives and the fruit trees start to bloom, the bees are already present and ready to begin their work.

One of the most common native pollinators of apples and other tree fruits are mason bees (*Osmia* spp.), and they are far more efficient at pollinating trees than honeybees are. This large group of bees includes approximately 150 different North American species, with the blue orchard mason bee (*O. lignaria*) being among the most important pollinators of fruit crops. Mason bees are tunnel nesters that create brood tunnels in wood and hollow stems. They can be very abundant and are important early-season pollinators. They carry pollen on their bellies instead of on their hind legs as many other bees do, so their abdomen is often coated with whatever color pollen they were feeding on. Mason bees use mud or chewed-up leaves to separate the brood cells in their nests. They are typically between ¼ to ¾ inch long and usually a blue-black color, sometimes with gray hairs. Some species are iridescent blue or green. Mason bees are sold commercially as cocoons in paper tubes for hanging near fruit trees.

Mason bee nest tube

Having a wide variety of flowering plants ensures that native bees will already be present when fruit trees come into bloom.

If you choose to purchase mason bees, be sure they were reared in your own region of the country. Many diseases and unwanted genetic traits are spread by transporting bees into other regions. It may prove healthier for the bees if you instead encourage the indigenous population by setting up an empty nesting block or bundles of hollow twigs and stems to encourage the indigenous population to take up residence.

Mining bees (*Andrena* spp.) are another group of bees that pollinate tree fruits, including cherries, plums, peaches, and more. There are more than four hundred species of mining bees in North America, and they're particularly abundant in spring. Measuring between ¼ and ½ inch long, most species are dark in color with no stripes on their abdomen. They're solitary, but multiple females often excavate brood chambers close to each other to form an in-ground colony in an area of exposed, sandy soil.

Native bees pollinate a large diversity of flowering plants and trees.

To encourage these native bees in orchards, partner your tree fruits with a diversity of blooming plants that provide nectar and pollen from early spring straight through to autumn.

Plants in the rose family (Rosaceae) are good choices for early nectar:

- Geums (*Geum* spp.)
- Serviceberry (*Amelanchier* spp.)
- Shrubby cinquefoil (*Potentilla* spp.)
- Smooth rose (*Rosa blanda*)
- Spireas (*Spiraea* spp.)
- Sweet briar rose (*Rosa rubiginosa*)

Other early bloomers include

- American basswood (*Tilia americana*)
- Blackberries (*Rubus* spp.)
- Dandelions (*Taraxacum officinale*)
- Ninebark (*Physocarpus* spp.)
- Red maple (*Acer rubrum*)
- Sumac (*Rhus* spp.)
- Willows (*Salix* spp.)

Late-season nectar and pollen sources include

- Asters (*Symphyotrichum* spp.)
- Coneflowers (*Echinacea* spp.)
- Goldenrods (*Solidago* spp.)
- Milkweeds (*Asclepias* spp.)
- Penstemons (*Penstemon* spp.)
- Tickseeds (*Coreopsis* spp.)

CREATE NESTING HABITAT *for* POLLINATORS

In the same way that companion planting can provide habitat for pest-eating beneficial insects (see page 170), it can also provide habitat for native pollinators. Of the 4,000 or so species of native bees in North America, about 1,200 nest in aboveground tubes and tunnels rather than building their brood chambers in the ground. Yes, you can purchase or build nesting tubes, tunnels, and blocks for these bees, but there's evidence that these artificial nesting sites cause a buildup of pathogens and parasites and may increase predation by birds and predatory insects.

Instead, consider planting plenty of hollow-stemmed plants, such as elderberries (*Sambucus* spp.), box elder (*Acer negundo*), joe-pye weed (*Eutrochium* spp.), brambles (*Rubus* spp.), cup plant (*Silphium* spp.), and bee balm (*Monarda* spp.), for these little bees to naturally nest in. While I may be stretching it a little to call this companion planting, I think it's just that. You're partnering hollow-stemmed plants with vegetable garden plants or fruit trees with the purpose of improving pollination rates — a pretty clear partnership and benefit in my eyes!

The remaining bee species build their brood chambers and take shelter in underground tunnels. They, too, deserve a bit of attention from gardeners when it comes to the preservation of nesting habitat.

A native bee crawls into a hollow stem to use it as a brood chamber.

Here are some ways you can help both tunnel-nesting and ground-nesting bee species:

- Be smart about garden cleanups. Do not cut back and clean up perennials and ornamental grasses in autumn. Instead, delay your garden cleanup until spring, when temperatures rise into the 50s (10° to 15°C) and pollinators have emerged.
- When cutting back your garden in spring, leave 8 to 12 inches of stubble behind. Many tunnel-nesting bees will make a home in these old plant stems.
- Eliminate pesticide use. And since the foraging ranges of native bees extend anywhere from 50 feet to half a mile, depending on the species, try to convince your neighbors and community to restrict pesticide use, too.
- If they aren't a hazard, don't remove snags (dead trees). Also, leave some brush piles and rock mounds in place. These areas can serve as nesting sites.
- Bare ground is essential for ground-nesting bee species. Don't cover every last inch of your property with mulch. Allow areas of exposed soil to remain undisturbed. Many native bees prefer to nest in south-facing slopes.
- Go no-till in the vegetable garden to preserve existing ground-based nesting sites and encourage the establishment of new ones.

A mining bee makes its home in a small nest cavity in the ground.

Once Again: Diversify!

One final way to improve pollination through companion planting is to simply increase the number of varieties you grow. The goal in conserving bees of all types is to make sure they have food available to them through the whole season. Growing a wide variety of fruits and vegetables can be part of the solution. Tree fruit, raspberries, peppers, strawberries, and a broad diversity of vegetable crops provide a season-long wave of blooms and lots of available nectar, especially when you throw plenty of other flowering plants into the mix. Plus, you'll extend your harvest and have more homegrown fruits and veggies to savor.

I'm sure by now you've come to see that a huge part of an effective companion planting strategy involves increasing the overall diversity of your garden rather than simply combining one specific plant with another. It's a theme I hope I've hammered home throughout the pages of this book, regardless of whether you're using companion planting to tackle pests, improve the soil, enhance pollination, or reduce weeds and diseases.

Epilogue

THE COMPANION PLANTING JOURNEY

In truth, like many aspects of gardening, companion planting is a journey. It isn't a cookie-cutter "fix" to problems. Rather, it's a road filled with trial and error, experimentation, and plenty of observation and note taking. It requires lots of flexibility on the part of the gardener, and the knowledge and acceptance that what works in one location, or in one particular season, may not be as successful in another. However, when it comes to the overarching ability of companion planting to increase the diversity of the garden and bring stability to the space, there's no doubt there is much to be gained, regardless of how much measurable success is visible to the gardener's eye.

Even as you seek specific results in terms of pests controlled, soil improved, and diseases suppressed, I suspect that you'll find a far greater gift on your companion planting journey. You'll discover the true value of thinking about your garden as an ecosystem rather than as a contrived environment. The small creatures that call your garden home will soon be seen for what they are: invaluable cogs in the workings of an amazing web of life.

In this world of diminishing habitat for plants and animals of all sorts, our gardens can become a safe harbor for creatures far smaller — and yet in many ways, far more important — than the humans who tend it. As you work to achieve a balance within the garden that benefits both humans and nonhumans alike, my hope is that your eyes will be opened to the value of gardens as wild spaces, where diverse flora and fauna are appreciated, understood, and even nurtured for the roles they play, regardless of a role's perceived value to human beings.

Though vegetable gardens, by their very nature, are cultivated, artificial environments, you have the opportunity to use that space for so much more than simply growing food: You can grow nature, both above and below the ground. You can improve the biodiversity of your neighborhood and community. You can provide much-needed resources for creatures whose natural habitat is disappearing. And you can educate others. Use the companion planting strategies you employ to show visitors the value of plant partnerships, whether specific or general. Use them to teach both adults and children about plants, pollinators, soil health, and even the fascinating lifecycles of pests. Both you, and your garden, will be better for it. Plus, you'll get tomatoes.

ACKNOWLEDGMENTS

Writing a book is a humbling experience, filled with lots of "alone time," research, and organizing. It begins long before a single word hits the page and continues for months after the manuscript is turned in. Though most folks consider the writer to be the most critical element in creating a book, nothing could be further from the truth. Were it not for a talented crew of smart and fantastic people behind this book, you wouldn't be holding it in your hands right now. I'm grateful to so many for their help throughout the process.

To Carleen Madigan, my acquisitions editor: thank you for trusting me with this topic and having enough faith to let me run with it. It has been such a pleasure working with you!

Hannah Fries, Michaela Jebb, and the rest of the incredible crew at Storey Publishing, I can't tell you how much I appreciate your attention to detail, your excellent communication skills, and your good nature. From editing to page design and everything in between, you have been nothing but remarkable.

My deepest appreciation to Kelly Smith Trimble for growing many of the plant pairings featured in the book, and to her husband, Derek, for providing the spectacular photos of those combos. Thank you for the care you gave to each companion planting technique; the results are so lovely!

Another big thank you to Holly Jones and the University of Tennessee Gardens in Knoxville for growing many of the plant partnerships featured in the book and allowing them to be photographed.

I extend a gracious tip of the hat to my new friend Angelo Merendino for his repeated travels to my home in Pittsburgh to photograph some of the plant partners growing in my own garden. Angelo, my family enjoyed meeting you, dining with you, and hearing about your travel and music adventures. Come back for more mojitos . . . and bring your sweetheart!

To Dr. Jeff Gillman: Thank you for the fantastic foreword you wrote for *Plant Partners* and for the amazing work you do educating the public about the science of horticulture. Readers, if you haven't read Jeff's books yet, I suggest getting started with *The Truth about Garden Remedies*. Thank you, Jeff!

And lastly, to my husband, John, and son, Ty: Thanks for loving me in all my plant nerd glory.

— *Jessica*

Resources

While there are dozens of seed catalogs and companies from which to source seeds for popular vegetable crops, sources for cover crop seeds and living mulch seeds can be more challenging to find. Here are some of my favorites.

American Meadows
2438 Shelburne Road, Suite 1
Shelburne, VT 05482
877-309-7333
www.americanmeadows.com

Arrow Seed
PO Box 722
Broken Bow, NE 68822
800-622-4727
www.arrowseed.com

Fedco Seeds
PO Box 520
Clinton, ME 04927
207-426-9900
www.fedcoseeds.com

Hearne Seed
512 Metz Road
King City, CA 93930
800-253-7346
www.hearneseed.com

High Mowing Seeds
76 Quarry Road
Wolcott, VT 05680
866-735-4454
www.highmowingseeds.com

Johnny's Selected Seeds
PO Box 299
Waterville, ME 04903
877-564-6697
www.johnnyseeds.com

Peaceful Valley Farm Supply
125 Clydesdale Court
Grass Valley, CA 95945
888-784-1722
www.groworganic.com

Southern Exposure Seed Exchange
PO Box 460
Mineral, VA 23117
540-894-9480
www.southernexposure.com

Sustainable Seed Company
175 West 2700 South
South Salt Lake, UT 84115
866-948-4727
www.sustainableseedco.com

Territorial Seed Company
PO Box 158
Cottage Grove, OR 97424
800-626-0866
www.territorialseed.com

West Coast Seeds
5300 34B Avenue
Delta, BC, Canada, V4L 2P1
888-804-8820
www.westcoastseeds.com

Glossary

allelochemicals. Chemicals produced and released by plants to inhibit the growth of other plants.

allelopathy. The ability of one plant to produce and release chemicals that inhibit the growth of other plants; a form of chemical competition.

appropriate/inappropriate landings theory. The theory that insects must make a set number of landings on the correct host plant to receive a strong enough cue to initiate egg-laying behavior.

banker plants. Sacrificial companion plants intentionally grown to attract and support pests so that beneficial insects can use them as a food source when pest populations are low in crop plants.

beetle banks. Long, raised berms of soil planted densely with native bunchgrasses to serve as habitat for pest-eating ground beetles. On a smaller scale, they are known as beetle bumps.

biodrilling. A process that uses cover crops to break up compacted soil to improve water filtration and gas exchange.

biological control or biocontrol. The science of using one living organism to help control the population of another. Augmentative biological control involves intentional release of the organism on a small scale. Classic biological control refers to intentional release on a large, outdoor scale. Conservation biological control refers to protecting and promoting the beneficial organisms that already exist in the landscape.

buzz pollination. Plant pollination that takes place when certain species of bees vibrate their flight muscles quickly, loosening pollen held firmly by the anther.

companion planting. The close pairing of two or more plant species for the benefit of one or more of them.

cover crops. Nonharvested crops planted either before or after the harvest of a vegetable crop, or in fallow fields and gardens.

crop rotation. Different crops planted in different locations or at different times of the year; a form of intercropping/interplanting.

diffusion. The movement of particles from an area of higher concentration to an area of lesser concentration.

diversity. The range of different species present in an environment.

ectomycorrhizae. Fungi whose hyphae (roots) do not penetrate plant cells and are found primarily in the soil surrounding plant roots.

endomycorrhizae. Fungi whose hyphae (roots) penetrate plant cells.

exudates. See root exudates.

friability. The ease with which soil crumbles and fragments from larger clods into smaller ones.

fungistasis. The inhibition of fungal organisms.

fusarium wilt. A disease that restricts the movement of water throughout an affected plant, causing branches or vines to turn yellow and wilt; eventually the entire plant dies.

green manure. Cover crops that are turned into the soil to add nutrients and organic matter, and to improve the soil's structure.

herbivore-induced plant volatiles. Chemical signals emitted by pest-infested plants. Also known as green leaf volatiles. See also semiochemicals.

hyphae. Threadlike fungal roots; see also mycorrhizae.

intercropping. The practice of growing multiple crops in the same area to promote beneficial results. On the smaller scale of the home garden, it's called interplanting.

mixed intercropping/interplanting. Partner crop plants blended together with no distinct rows or other formal arrangement.

mycorrhizae. Fungi that colonize the roots of plants, primarily in a symbiotic relationship.

nitrogen fixation. The process by which certain soil microorganisms take atmospheric nitrogen and convert it into a form plants can use.

parasitoidal insects. Those that use other insects to house and feed their developing young, eventually killing the host insect in the process.

pathogen. A virus or microorganism that causes disease.

polyculture. An agriculture system in which multiple plants are growing in the same space to reflect the diversity of a natural ecosystem and avoid the presence of a monoculture where pests and diseases can easily spread from one plant to the next.

predatory insects. Those that capture and consume other insects directly.

push-pull system. An insect pest trapping system that combines the vegetable crop with a trap crop that lures the pests away and another plant that masks or hides the vegetable crop.

relay intercropping. On farms, when a second crop is planted right into an existing crop just before harvest of the first crop.

resource concentration hypothesis. The theory that plant-eating insects are less likely to find host plants in diverse habitats and more likely to find them when host plants are found in higher densities.

rhizosphere. The area around a plant root.

root exudates. Sugars, amino acids, enzymes, vitamins, and many other compounds produced and excreted by the roots of living plants.

row intercropping/interplanting. Mixed crops grown together in alternating rows.

semiochemicals. Chemical signals released by plants and insects into the air or soil to modify the behavior of another organism.

specific disease suppression. An increase in particular soil microbes that act directly against pathogenic microorganisms.

structural complexity. The assortment of growth habits and structures of a diverse group of plants.

trap cropping. A companion planting technique for pest management where veggies are planted in conjunction with "sacrificial" companion plants whose economic benefit comes solely from their ability to draw pests away from the vegetables to be harvested.

verticillium wilt. A fungal disease found in soils that affects the water-conducting tissue of its host, causing the plant to yellow, wilt, and usually die prematurely.

METRIC CONVERSIONS

TO CONVERT	TO	MULTIPLY
pounds	kilograms	pounds by 0.45
inches	centimeters	inches by 2.54
feet	meters	feet by 0.3048

Bibliography

Chapter 1:
The Power of Plant Partnerships

Andow, D. A. (1991). "Vegetational diversity and arthropod population response." *Annual Review of Entomology, 36,* 561–586.

Beyfuss, R. (1994). "Companion Planting, Ecogardening Fact Sheet #10." Cornell University.

Blande, J. D., Pickett, J. A., & Poppy, G. M. (2007). "A comparison of semiochemically mediated interactions involving specialist and generalist Brassica-feeding aphids and the braconid parasitoid Diaeretiella rapae." *J. Chem. Ecol., 33,* 767–779.

Bruce, T. J. A., Wadhams, L. J., & Woodcock, C. M. (2005). "Insect host location: a volatile situation." *Trends Plant Sci., 10,* 269–274.

Buranday, R. P. & Raros, R. S. (1975). "Effects of cabbage-tomato intercropping on the incidence and oviposition of the diamond back moth, Plutella xylostella (L.)." *Phillippine Entomology, 2,* 369–375.

De Moraes, C. M., Lewis, W. J., Pare, P. W., Alborn, H. T., & Tumlinson, J. H. (1998). "Herbivore-infested plants selectively attract parasitoids." *Nature, 393,* 570–573.

Du, Y. J., Poppy, G. M., & Powell, W. (1996). "Relative importance of semiochemicals from first and second trophic levels in host foraging behavior of Aphidius ervi." *J. Chem. Ecol., 22,* 1591–1605.

Du, Y. J., Poppy, G. M., Powell, W., Pickett, J. A., Wadhams, L. J., & Woodcock C. M. (1998). "Identification of semiochemicals released during aphid feeding that attract parasitoid Aphidius ervi." *J. Chem. Ecol., 24,* 1355–1368.

Elhakeem, A., Markovic, D., Broberg A., Anten, N.P.R., & Ninkovic, V. (2018). "Aboveground mechanical stimuli affect belowground plant-plant communication." *PLoS ONE, 13*(5), e0195646.

Farmer, E. E. & Ryan, C. A. (1990). "Interplant communication: Airborne methyl jasmonate induces synthesis of proteinase inhibitors in plant leaves." *Proc. Natl. Acad. Sci. USA, 87,* 7713–7716.

Feeny, P. (1976). "Plant apparency and chemical defense." Biochemical interaction between Plants and Insects (Wallace, J. & Mansell, R., eds.)," *Recent Advances in Phytochemistry,* 1–40. Springer.

Gao, Z. Z., Wu, W. J., & Cui, Z. X. (2004). "The Effect of Intercrop on the Densities of Phyllotreta Striolata (F.)." *Chinese Agricultural Science Bulletin, 20,* 214–216.

Gilbert, L. & Johnson, D. (2017). "Plant–plant communication through common mycorrhizal networks." *Advances in Botanical Research, 82,* 83–97.

Girling, R. D., Stewart-Jones, A., Dherbecourt, J., Staley, J. T., Wright, D. J., & Poppy, G. M. (2011). "Parasitoids select plants more heavily infested with their caterpillar hosts: a new approach to aid interpretation of plant headspace volatiles." *Proceedings of the Royal Society B: Biological Sciences, 278*(1718), 2646–2653.

Gorzelak, M. A., Asay, A. K., Pickles, B. J., & Simard, S. W. (2015). "Inter-plant communication through mycorrhizal networks mediates complex adaptive behaviour in plant communities." *AoB Plants, 7.*

Heil, M., & Karban, R. (2010). "Explaining evolution of plant communication by airborne signals." *Trends Ecol. Evol., 25,* 137–144.

Heil, M., & Silva Bueno, J. C. (2007). "Within-plant signaling by volatiles leads to induction and priming of an indirect plant defense in nature." *Sci. Signal. 104,* 5467.

Huang, W., Gfeller, V., & Erb, M. (2019). "Root volatiles in plant–plant interactions II: Root volatiles alter root chemistry and plant–herbivore interactions of neighbouring plants." *Plant, Cell & Environment, 42*(6), 1964–1973.

Hunter, A. & Aarssen, L. (1988). "Plants helping plants." *BioScience, 38,* 34–40.

Jabran, K. (2017). "Rye allelopathy for weed control." *Manipulation of Allelopathic Crops for Weed Control,* 49–56. Springer.

Karban, R., Baldwin, I., Baxter, K., Laue, G., & Felton, G. (2000). "Communication between plants: Induced resistance in wild tobacco plants following clipping of neighboring sagebrush." *Oecologia, 125,* 66–71.

Lithourgidis, A., Dordas, C., Damalas, C., & Vlachostergios, D. (2011). "Annual intercrops: An alternative pathway for sustainable agriculture." *Aust. J. Crop Sci., 5,* 396.

Meier, Amanda R. & Hunter, Mark D. (2018). "Arbuscular mycorrhizal fungi mediate herbivore-induction of plant defenses differently above and belowground." *Oikos, 127*(12), 1759–1775.

Novoplansky, A. (2009). "Picking battles wisely: Plant behaviour under competition." *Plant Cell Environ., 32,* 726–741.

Oelmüller, R. (2019). "Interplant communication via hyphal networks." *Plant Physiology Reports,* 1–11.

Parker, J. E., Snyder, W. E., Hamilton, G. C., & Rodriguez-Saona, C. (2013). "Companion planting and insect pest control." *Weed and Pest Control—Conventional and New Challenges* (Soloneski & S., Larramendy, M., eds.), 1–30. InTech.

Perrin, R. M. & Phillips, M. L. (1978). "Some effects of mixed cropping on the population dynamics of insect pests." Entomol. Exp. .Appli., *24,* 385–393.

Pieterse, C. M. J. & Dicke, M. (2007). "Plant interactions with microbes insects: from molecular mechanisms to ecology." *Trends Plant Sci., 12,* 564–569.

Simard, S. W. (2018). "Mycorrhizal networks facilitate tree communication, learning, and memory." *Memory and Learning in Plants* (Baluska, F., Gagliano, M., & Witzany, G., eds.), 191–213. Springer.

Song, Y., Wang, M., Zeng, R., Groten, K., & Baldwin, I. T. (2019). "Priming and filtering of antiherbivore defenses among plants connected by mycorrhizal networks." *Plant, Cell & Environment, 42*(11), 2945–2961.

Tahvanainen, J. O. & Root, R. B. (1972). "The influence of vegetational diversity on population ecology of a specialize herbivore, Phyllotreta cruciferae (Coleoptera: Chrysomelidae)." *Oecologia, 10,* 321–346.

Turlings, T. C. J., Tumlinson, J. H., & Lewis, W. J. (1990). "Exploitation of herbivore-induced plant odors by host-seeking parasitic wasps." *Science, 250,* 1251–1253.

Turlings, T. C. J., Wackers, F. L.; Vet, L. E. M.; Lewis, W. J. & Tumlinson, J. H. (1993). "Learning of host-finding cues by hymenopterous parasitoids." *Insect learning: ecological and evolutionary perspectives* (Papaj, D. R. & Lewis, A. C., eds.), 51–78. Chapman and Hall, Inc.

Vandermeer, J. H. (1992). *The Ecology of Intercropping.* Cambridge University Press.

Vet, L. E. M. & Dicke, M. (1992). "Ecology of infochemical use by natural enemies in a tritrophic context." *Annu. Rev. Entomol. 37,* 141–172.

Chapter 2: Soil Preparation & Conditioning

ATTRA. (2006). "Overview of Cover Crops and Green Manures." http://attra.ncat.org/attra-pub/PDF/covercrop.pdf

Chen, G. & Weil, R. R. (2010). "Penetration of cover crop roots through compacted soils." *Plant and Soil, 331*(1–2), 31–43.

Costello, M. J. (1994). "Broccoli growth, yield and level of aphid infestation in leguminous living mulches." *Biol. Ag. and Hort., 10,* 207–222.

Flexner, J. L. (1990). Hairy vetch. Univ. of Calif. SAREP Cover Crops Resource Page. www.sarep.ucdavis.edu/ccrop

Fox, R. H. & Piekielek, W. P. (1988). "Fertilizer N Equivalence of Alfalfa, Birdsfoot Trefoil, and Red Clover for Succeeding Corn Crops." *J. Prod. Agric., 1,* 313–317.

Gruver, J., Weil, R., White, C., & Lawley, Y. (2014). *Radishes: a new cover crop for organic farming systems.* Michigan State University.

He, X., Critchley, C., & Bledsoe, C. (2003). "Nitrogen Transfer Within and Between Plants Through Common Mycorrhizal Networks (CMNs)." *Critical Reviews in Plant Sciences, 22*(6), 531–567.

Holmes, A. (2016). "Soil Health and the Tillage Radish." http://hdl.handle.net/2142/89900

House, G. J. & Alzugaray, M. D. R. (1989). Hairy vetch. Univ. of Calif. SAREP Cover Crops Resource Page. www.sarep.ucdavis.edu/ccrop

Kahn, B. A. (2010). "Intercropping for field production of peppers." *HortTechnology, 20*(3), 530–532.

Knight, W. E. (1985). Crimson clover. Univ. of Calif. SAREP Cover Crops Resource Page. www.sarep.ucdavis.edu/ccrop

Larkin, R. P., Griffin, T. S., & Honeycutt, C. W. (2010). "Rotation and cover crop effects on soilborne potato diseases, tuber yield, and soil microbial communities." *Plant Disease, 94*(12), 1491–1502.

Li, L., Yang, S., Li, X., Zhang, F. & Christie, P. (1999). "Interspecific complementary and competitive interactions between intercropped maize and faba bean." *Plant and Soil, 212*(2), 105–114.

Marks, C. F. & Townsend, J. L. (1973). Buckwheat. Univ. of Calif. SAREP Cover Crops Resource Page. www.sarep.ucdavis.edu/ccrop

Michigan State Univ. Extension. Cover Crops Program. East Lansing, Mich. www.covercrops.msu.edu

Montesinos-Navarro, A., Verdú, Miguel, Ignacio Querejeta, J., Sortibrán, L., & Valiente-Banuet, A. (2016). "Soil fungi promote nitrogen transfer among plants involved in long-lasting facilitative interactions." *Perspectives in Plant Ecology, Evolution and Systematics*, 10.1016/j.ppees.2016.01.004.

Phatak, S. C., et al. (1991). "Cover crops effects on weeds diseases, and insects of vegetables." *Cover Crops for Clean Water* (Hargrove, W. L., ed.), 153–154. Soil and Water Conservation Society.

Chapter 3: Weed Management

Abdul-Baki, A. A. & Teasdale, J. R. (1993). "A no-tillage tomato production system using hairy vetch and subterranean clover mulches." *HortSci., 28,* 106–108.

Adhikari, L., Mohseni-Moghadam, M., & Missaoui, A. (2018). "Allelopathic Effects of Cereal Rye on Weed Suppression and Forage Yield in Alfalfa." *American Journal of Plant Sciences, 9,* 685–700.

Biazzo, J. & Masiunas., J. B. (2000). "The use of living mulches for weed management in hot pepper and okra." *J. Sustainable Agr., 16,* 59–79.

Boydston, R. A. & Al-Khatib, K. (2005). "Utilizing Brassica cover crops for weed suppression in annual cropping systems." *Handbook of Sustainable Weed Management* (Singh, H. P., Batish, D. R., & Kohli, R. K., eds.), 77–94. Haworth Press.

Broughton, S. E. (2010). "The Effects of Living Mulches on Organic, Reduced-Till Broccoli Growth and Management." Master's Thesis, University of Tennessee.

Costello, M. J. & Altieri, M.A. (1995). "Abundance, growth rate and parasitism of *Brevicoryne brassicae* and *Myzus persicae* (Homoptera: Aphididae) on broccoli grown in living mulches." *Agr. Ecosystems Environ., 52,* 187–196.

Chon, S. U. & Nelson, C. (2010). "Allelopathy in Compositae plants. A review." *Agron. Sustain. Dev. 30,* 349–358.

Farooq, M., Jabran, K., Cheema, Z. A., Wahid, A., & Siddique, K. H. M. (2011). "The role of allelopathy in agricultural pest management." *Pest Manag. Sci., 67,* 493–506.

Fisk, J. W., et al. (2001). "Weed suppression by annual legume cover crops in no-tillage corn." *Agron. J., 93,* 319–325.

Fujiyoshi, Phillip. (1998). "Mechanisms of Weed Suppression by Squash (Curcurbita spp.) Intercropped in Corn (Zea mays)." Dissertation, University of California Santa Cruz.

Gardiner, J. B. et al. (1999). "Allelochemicals released in soil following incorporation of rapeseed (Brassica napus) green manures." *J. Agric. Food Chem., 47,* 3837–3842.

Haramoto, E. R. & Gallandt., E. R. (2005). "Brassica cover cropping: I. Effects on weed and crop establishment." *Weed Sci., 53,* 695–701.

Hartwig, N. L. & Ammon, H. U. (2002). "Cover crops and living mulches." *Weed Sci., 50,* 688–699.

Hooks, C. R. R., Valenzuela, H. R. & DeFrank, J. (1998). "Incidence of pests and arthropod natural enemies in zucchini grown with living mulches." *Agriculture, Ecosystems and Environment, 69,* 217–231.

Hooks, C. R. & Johnson, M. W. (2006). "Population densities of herbivorous lepidopterans in diverse cruciferous cropping habitats: effects of mixed cropping and using a living mulch." *BioControl, 51,* 485–506.

Ilnicki, R. D. & Enache, A. J. (1992). "Subterranean clover living mulch: an alternative method of weed control." *Agr., Ecosystems Environ., 40,* 249–264.

Jabran, K. (2017). "Rye allelopathy for weed control." *Manipulation of Allelopathic Crops for Weed Control,* 49–56. Springer International Publishing AG.

Kelly, T. C., et al. (1995). "Economics of a hairy vetch mulch system for producing fresh-market tomatoes in the Mid-Atlantic region." *HortSci., 120,* 854–869.

Mohler, C. L. (1995). "A living mulch (white clover) / dead mulch (compost) weed control system for winter squash." Proc. Northeast. Weed *Sci. Soc., 49,* 5–10.

Paine, L., et al. (1995). "Establishment of asparagus with living mulch." *J. Prod. Agric. 8,* 35–40.

Rice, E. L. (1984). *Allelopathy.* Academic Press.

Singh, H. P., Batish, D. R., & Kohli, R. K. (2001). "Allelopathy in agro-ecosystems." *J. CropProd., 4,* 1–41.

Theriault, F., Stewart, K. A, & Seguin, P. (2009). "Use of perennial legumes living mulches and green manures for the fertilization of organic broccoli." *Intl. J. Veg. Sci, 15,* 142–157.

Theunissen, J., Booij, C. J. H., Schelling, G., & Noorlander, J. (1992). "Intercropping white cabbage with clover." IOBC/WPRS Bulletin XV, 4, 104–114.

Trinchera, A., et al. (2019). "Mycorrhiza-mediated interference between cover crop and weed in organic winter cereal agroecosystems: The mycorrhizal colonization intensity indicator." *Ecology and Evolution, 9*(10), 5593–5604.

Wiles, L. J., William, R. D., Crabtree, G. D., & Radosevish, S. R. (1989). "Analyzing competition between a living mulch and a vegetable crop in an interplanting system." *J. Amer. Soc. Hort. Sci. 115,* 1029–1034.

Chapter 4: Support & Structure

No references.

Chapter 5: Pest Management

Amarawardana, L., Bandara, P., Kumar, V., Pettersson, J., Ninkovic, V., & Glinwood, R. (2007). "Olfactory response of *Myzus persicae* (Hemiptera: Aphididae) to volatiles from leek and chive: Potential for intercropping with sweet pepper." *Acta Agric. Scand. B, 57,* 87–91.

Andersson, M. (2007). "The effects of non-host volatiles on habitat location by phytophagous insects." Introductory Paper at the Faculty of Landscape Planning, Horticulture and Agricultural Science, Swedish University of Agricultural Sciences, 1–38. Alnap.

Andow, D. (1991). "Vegetational diversity and arthropod population response." *Annual Review of Entomology, 36,* 561–586.

Arimura, G. I., Ozawa, R., Shimoda, T., Nishioka, T., Boland, W., & Takabayashi, J. (2000). "Herbivory-induced volatiles elicit defense genes in lima bean leaves." *Nature, 406,* 512–515.

Baldwin, I. T., Kessler, A., & Halitschke, R. (2002). "Volatile signaling in plant-plant-herbivore interactions: What is real?" *Curr. Opin. Plant Biol., 5,* 351–354.

Ben-Issa, R., Gomez, L., & Gautier, H. (2017). "Companion Plants for Aphid Pest Management." Institut National de Recherche Agronomique, France.

Boucher, T. J. & Durgy, R. (2004). "Moving towards ecologically based pest management: a case study using perimeter trap cropping." *Journal of Extension, 42*(6).

Brooker, R. W., et al. (2015). "Improving intercropping: A synthesis of research in agronomy, plant physiology and ecology." *New Phytol., 206,* 107–117.

Bugg, R. L., Chaney, W. E., Colfer, R. G., Cannon, J. A., & Smith, H. A. (2008). "Flower flies (Diptera: Syrphidae) and other important allies in controlling pests of California vegetable crops." University of California, Division of Agriculture and Natural Resources, Publication 8285. University of California Press.

Collier, R. H. & Finch, S. (2003). "The Effect of Increased Crop Diversity on Colonisation by Pest Insects of Brassica Crops." *Crop Science and Technology*, 439–444. British Crop Protection Council.

Döring, T. F. (2014). "How aphids find their host plants, and how they don't." *Ann. Appl. Biol., 165*, 3–26.

Dover, J. W. (1986). "The effects of labiate herbs and white clover on Plutella xylostella oviposition." *Entomol. Exp. Appl., 42*(3), 243–247.

Finch, S. & Collier, R. (2000). "Host-plant selection by insects — a theory based on appropriate/inappropriate landings by pest insects of cruciferous plants." *Entomol. Exp. Appl., 96*, 91–102.

Finch, S., Billiald, H., & Collier, R. H. (2003). "Companion planting — Do aromatic plants disrupt host-plant finding by the cabbage root fly and the onion fly more effectively than non-aromatic plants?" *Entomol. Exp. Appl., 109*(3), 183–195.

Hagler, J. R., Nieto, D. J., Machtley, S. A, Spurgeon, D. W., Hogg, B. N., & Swezey, S. L. (2018). "Dynamics of Predation on Lygus hesperus (Hemiptera: Miridae) in Alfalfa Trap-Cropped Organic Strawberry." *Journal of Insect Science, 18*(4), 12.

Hokkanen, H. M. T. (1991). "Trap cropping in pest management." *Annu. Rev. Entomol., 36*, 119–138.

Hooks, C. R. R. & Johnson, M. W. (2003). "Impact of agricultural diversification on the insect community of cruciferous crops." *Crop Prot., 22*, 223–238.

Hurej, M. (2000). "Trap plants and their application in plant protection against pests." *Prog. Plant Prot., 40*, 249–253.

Jankowska, B., Poniedziaek, M., & Jedrszczyk, E. (2009). "Effect of intercropping white cabbage with French marigold (Tagetes patula nana L.) and pot marigold (Calendula officinalis l.) on the colonization of plants by pest insects." *Folia Hortic., 21*, 95–103.

Javaid, I. & Joshi, J. M. (1995). "Trap cropping in insect pest management." *Journal of Sustainable Agriculture*, 5(1–2), 117–136.

Koschier, E. H., Sedy, K. A. & Novak, J. (2002). "Influence of plant volatiles on feeding damage caused by the onion thrips (Thrips tabaci)." *Crop Protection, 21*(5), 419–425.

Kostal, V. & Finch, S. (1994). "Influence of background on host plant selection and subsequent oviposition by the cabbage root fly (Delia radicum)." *Entomol. Exp. Appl., 70*, 153–163.

Lopes, T., Hatt, S., Xu, Q., Chen, J., Liu, Y., & Francis, F. (2016). "Wheat (Triticum aestivum L.)-based intercropping systems for biological pest control." *Pest Manag. Sci., 72*, 2193–2202.

Ludwig, S. W. & Kok, L. T. (1998). "Evaluation of trap crops to manage harlequin bugs, Murgantia histrionica (Hahn)(Hemiptera: Pentatomidae) on broccoli." *Crop protection, 17*(2), 123–128.

Moreno, C. R. & Racelis, A. E. (2015). "Attraction, repellence, and predation: Role of companion plants in regulating Myzus persicae (Sulzer) (Hemiptera: Aphidae) in organic kale systems of south Texas." *Southwest. Entomol., 40*, 1–14.

Morley, K., Finch, S., & Collier, R. H. (2005). "Companion planting—behavior of the cabbage root fly on host plants and non-host plants." *Entomol. Exp. Appl., 117*, 15–25.

Pickett, J., Wadhams, L., & Woodcock, C. (1997). "Developing sustainable pest control from chemical ecology." *Agric. Ecosyst. Environ., 64*, 149–156.

Pollock, Sandra. Sustainable Agricultural Research & Education publication. "Trap Cropping in Vegetable Production: An IPM Approach to Managing Pests." https://ipm.ifas.ufl.edu/pdfs/trapcropsfactsheet.pdf

Potts, M. J. & Gunadi, N. (1991). "The influence of intercropping with allium on some insect populations in potato (Solatium tuberosum)." *Ann. Appl. Biol., 119*, 207–213.

Reddy, P. P. (2017). "Intercropping." *Agro-ecological Approaches to Pest Management for Sustainable Agriculture*, 109–131. Springer.

Reisselman, Leah. *Companion Planting: A Method for Sustainable Pest Control.* Iowa State University, Armstrong and Neely-Kinyon research and Demonstration Farm; RFR-A9099; ISRF09-12

Root, R. B. (1973). "Organization of a plant-arthropod association in simple and diverse habitats: the fauna of collards (Brassica oleracea)." *Ecol. Monog., 43,* 95–124.

Shelton, A. & Badenes-Perez, F. (2006). "Concepts and applications of trap cropping in pest management." *Annu. Rev. Entomol., 51,* 285–308.

Smith, J. G. (1976). "Influence of crop backgrounds on aphids and other phytophagous insects on Brussels sprouts." *Annals of Applied Biology, 83,* 1–13.

Srinivasan, K. & Krishna Moorthy, P. (1992). "Development and Adoption of Integrated Pest Management for Major Pests of Cabbage Using Indian Mustard as a Trap Crop." *Proceedings of the second international workshop on management of Diamondback moth and other crucifer pests*, Tainan, Taiwan, 10–14 December, 1992, 511–521.

Tahvanainen, J. O. & Root, R. B. (1972). "The influence of vegetational diversity on the population ecology of a specialized herbivore, Phyllotreta cruciferae (Coleoptera: Chrysomelidae)." *Oecologia, 10,* 321–346.

Thiery, D. & Visser, J. H. (1986). "Masking of host plant odour in the olfactory orientation of the Colorado potato beetle." *Entomol. Exp. Appl., 41,* 165–172.

Thiery, D. & Visser, J. H. (1987). "Misleading the Colorado potato beetle with an odor blend." *Journal of Chemical Ecology, 13,* 1139–1146.

Uvah, I. & Coaker, T. (1984) "Effect of mixed cropping on some insect pests of carrots and onions." *Entomol. Exp. Appl., 36,* 159–167.

Visser, J. & Avé, D. (1978). "General green leaf volatiles in the olfactory orientation of the colorado beetle, Leptinotarsa decemlineata." *Entomol. Exp. Appl., 24,* 738–749.

Chapter 6: Disease Management

Abdul-Baki, A. A., Stommel, J. R., Watada, A. E., Teasdale, J. R., & Morse, R. D. (1996). "Hairy vetch mulch favorably impacts yield of processing tomatoes." *HortSci., 31,* 338–340.

Brown, P. D., & Morra, M. J. (1997). "Control of soil-borne plant pests using glucosinolate-containing plants." *Adv. Agron., 61,* 167–215.

Dupont, Tianna. (2015). "Reducing Soil Borne Diseases with Cover Crops. PennState Extension." https://extension.psu.edu/reducing-soil-borne-diseases-with-cover-crops

Everts, Kathryne & Himmelstein, Jennifer. (2015). "Fusarium wilt of watermelon: Towards sustainable management of a re-emerging plant disease." *Crop Protection, 73.*

Fereres, A. (2000). "Barrier crops as a cultural control measure of nonpersistently transmitted aphid-borne viruses." *Virus Research, 71,* 221–231.

Frank, D. L., & Liburd, O.E. (2005). "Effects of living and synthetic mulch on the population dynamics of whiteflies and aphids, their associated natural enemies and insect-transmitted plant diseases in zucchini." *Environmental Entomology 34,* 857–865.

Hao, J. J. & Subbarao, K. V. (2006). "Dynamics of lettuce drop incidence and Sclerotinia minor inoculum under varied crop rotations." *Plant Disease, 90,* 269–278.

Hartwig, N. L., & Ammon, H.U. (2002). "Cover crops and living mulches." *Weed Science, 50,* 688–699.

Hooks, C. R. R., & Fereres, A. (2006). "Protecting crops from non-persistently aphid-transmitted viruses: A review on the use of barrier plants as a management tool." *Virus Research 120,* 1–16.

Hooks, C. R. R., Fereres, A., & Wang, K. H. (2007). "Using protector plants to guard crops from aphid-born non-persistent viruses." University of Hawai'i at Mānoa, College of Tropical Agriculture and Human Resources, publication SCM-18.

Ilnicki, R., & Enache, A. (1992). "Subterranean clover living mulch: An alternative method of weed control." *Agriculture, Ecosystems & Environment, 40,* 249–264.

Ju Ding, Yao Sun, Chun Lan Xiao, Kai Shi, Yan Hong Zhou, Jing Quan Yu. (2007). "Physiological basis of different allelopathic reactions of cucumber and figleaf gourd plants to cinnamic acid." *Journal of Experimental Botany, 58*(13), 3765–3773.

Keinath, A. P., Hassell, R. L., Everts, K. L., & Zhou, X. G. (2010). "Cover crops of hybrid common vetch reduce Fusarium wilt of seedless watermelon in the eastern United States." *Plant Health Progress, 11*(1), 8.

Kruidhof, H. M., Bastiaans, L., & Kropff, M. J. (2009). "Cover crop residue management for optimizing weed control." *Plant and Soil, 318,* 169–184.

Larkin, R., Griffin, T., & Honeycutt, C. (2010). "Rotation and Cover Crop Effects on Soilborne Potato Diseases, Tuber Yield, and Soil Microbial Communities." *Plant Disease, 94.*

Lockerman, R. H., & Putnam, A. R. (1979). "Evaluation of allelopathic cucumbers (Cucumis sativus) as an aid to weed control." *Weed Science, 27*(1), 54–57.

Ochiai, N., Powelson, M. L., & Crowe, F. J., et al. (2000). "Green manure effects on soil quality in relation to suppression of Verticillium wilt of potatoes." *Biol. Fertil. Soils, 44,* 1013–1023.

Paine, L., Harrison, H., and Newenhouse, A. (1995). "Establishment of Asparagus with Living Mulch." *J. Prod. Agric. 8,* 35–40.

Patten, K., Nimr, G., & Neuendorff, E. (1990). "Evaluation of Living Mulch Systems for Rabbiteye Blueberry Production." *HortSci., 25.*

Thresh M. (1982). "Cropping practices and virus spread." *Annual Review of Phytopathology, 20,* 193–218.

Zhou, X. G., & Everts, K. L. (2004). "Suppression of Fusarium wilt of watermelon by soil amendment with hairy vetch." *Plant Disease, 88,* 1357–1365.

Chapter 7: Biological Control

Andow, D. A. (1991). "Vegetational diversity and arthropod population response." *Annu. Rev. Entomol., 36,* 561–586.

Baggen, L. R., Gurr, G. M., & Meats, A. (1999). "Flowers in tri-trophic systems: mechanism allowing selective exploitation by insect natural enemies for conservation biological control." *Entomol. Exp. Appl., 91*(1), 155–161.

Balmer, O., et al. (2014). "Wildflower companion plants increase pest parasitation and yield in cabbage fields: Experimental demonstration and call for caution." *Biological Control, 76,* 19–27.

Begum, M., Gurr, G. M., Wratten, S. D., Hedberg, P., & Nicol, H. I. (2004). "The effect of floral nectar on the grapevine leafroller parasitoid Trichogramma carverae." *International Journal of Ecology and Environmental Sciences, 30,* 3–12.

Beizhou, S., Jie, Z., Jinghui, H., Hongying, W., Yun, K., & Yuncong, Y. (2010). "Temporal dynamics of the arthropod community in pear orchards intercropped with aromatic plants." *Pest Manag. Sci., 67,* 1107–1114.

Beizhou, S., Jie, Z., Wiggins, N. L., Yuncong, Y., Guangbo, T., & Xusheng, S. (2012). "Intercropping with aromatic plants decreases herbivore abundance, species richness, and shifts arthropod community trophic structure." *Environmental Entomology, 41,* 872–879.

Bickerton, M. W. & Hamilton, G. C. (2012). "Effects of Intercropping with flowering plants on predation of Ostrinia nubilalis (Lepidoptera: Crambidae) eggs by generalist predators in bell peppers." *Environmental Entomology, 41,* 612–620.

Collins, K. L., Boatman, N. D., Wilcox, A., Holland, J. M., & Chaney, K. (2002). "Influence of beetle banks on cereal aphid predation in winter wheat." *Agriculture, Ecosystems and Environment, 93,* 337–350.

Cowgill, S. E., Wratten, S. D., & Sotherton N. W. (1993). "The effect of weeds on the numbers of hoverfly (Diptera: Syrphidae) adults and the distribution and composition of their eggs in winter wheat." *Annals of Applied Biology, 123*, 499–514.

Dong, M., Zhang, D., & Du, X. (2011). "The relationship between aphids and their natural enemies and their ecological management." *Acta Phytophylacica Sin., 38*, 327–332.

Ehler, L. (1998). "Conservation biological control: Past, present, and future." *Conservation Biological Control*, 1–8. Academic Press.

Finch, S. & Edmonds, G. H. (1994). "Undersowing cabbage crops with clover — the effects on pest insects, ground beetles and crop yield." IOBC/WPRS Bulletin, *17*(8), 159–167.

Finch, S. & Kienegger, M. A. (1997). "Behavioural study to help clarify how undersowing with clover affects host plant selection by pest insects of brassica crops." *Entomol. Exp. Appl.*, 84, 165–172.

Géneau, C. E., Wackers, F. L., Luka, H., Daniel, C., & Balmer, O. (2012). "Selective flowers to enhance biological control of cabbage pests by parasitoids." *Basic Appl. Ecol., 13*, 85–93.

Heimpel, G. E., & Jervis, M. A. (2005). "Does floral nectar improve biological control by parasitoids?" *Plant-provided Food and Plant-Carnivore Mutualism* (Wackers, F. L., Van Rijn, P. C. J., & Bruin, J., eds.) 267–304. Cambridge University Press.

Hooks, C. R., Pandey, R. R., & Johnson, M. W. (2007). "Using clovers as living mulches to boost yields, suppress pests, and augment spiders in a broccoli agroecosystem." Univ. Hawai'i Coop. Ext. Serv. Publ. IP-27.

Jervis, M. A., & Heimpel. G. E. (1986). "Phytophagy in insects as natural enemies" (Jervis, M. A., Ed.), Springer, Netherlands, 525–550.

Jervis, M. A., & Kidd, N. A. C. (1986). "Host-feeding strategies in hymenopteran parasitoids." *Biological Reviews, 61*, 395–434.

Jonsson M., Wratten, S. D., Landis, D. A., & Gurr, G. M. (2008). "Recent advances in conservation biological control of arthropods by arthropods." *Biological Control, 45*, 172–175.

Landis, D. A., Wratten, S. D., & Gurr, G. M. (2000). "Habitat management to conserve natural enemies of arthropod pests in agriculture." *Annu. Rev. Entomol. 45*, 175–201.

Lavandero, B. I., Wratten, S. D., Didham, R. K., & Gurr, G. (2006). "Increasing floral diversity for selective enhancement of biological control agents: A double-edged sward?" *Basic and Applied Ecology, 7*, 236–243.

Letourneau, D. K., & Altieri, M. A. (1999). "Environmental management to enhance biological control in agroecosystems." *Handbook of Biological Control* (Bellows, T.S., & Fischer, T.W., eds.), 319–354. Academic Press.

Lövei, G. L., Hodgson, D. J., MacLeod, A., & Wratten, S. D. (1993). "Attractiveness of some novel crops for flower-visiting hover flies (Diptera: Syrphidae): comparisons from two continents." *Pest control and sustainable agriculture* (Corey, S., Dall, D., & Milne, W., eds.), 368–370. CSIRO Publications.

Lu, Z. X., et al. (2014). "Mechanisms for flowering plants to benefit arthropod natural enemies of insect pests: Prospects for enhanced use in agriculture." *Insect Sci., 21*, 1–12.

MacLeod, A. (1992). "Alternative crops as floral resources for beneficial hoverflies (Diptera: Syphidae)." Proceedings of the Brighton Crop Protection Conference, 997–1002. British Crop Protection Council.

Maredia, K. M., Gage, S. H., Landis, D. A., & Scriber, J. M. (1992). "Habitat use patterns by the seven-spotted lady beetle (Coleoptera: Coccinellidae) in a diverse agricultural landscape." *Biological Control, 2*, 159–165.

Morandin, L. A., & Kremen, C. (2013). "Hedgerow restoration promotes pollinator populations and exports native bees to adjacent fields." *Ecol. Appl., 23*, 829–839.

Morris, M. C., & Li, F. Y. (2000) "Coriander (*Coriandrum sativum*) companion plants can attract hoverflies, and may reduce pest infestation in cabbages." *N. Z. J. Crop Hortic. Sci., 28*, 213–217.

Patt, J. M., Hamilton, G. C., & Lashomb, J. H. (1997) "Foraging success of parasitoid wasps on flowers: interplay of insect morphology, floral architecture and searching behavior." *Entomol. Exp. Appl., 83,* 21-30.

Perrin, R., & Phillips, M. (1978). "Some effects of mixed cropping on the population dynamics of insect pests." *Entomol. Exp. Appl., 24,* 585–593.

Potting, R. P. J., Poppy, G. M., & Schuler, T. H. (1999). "The role of volatiles from cruciferous plants and pre-flight experience in the foraging behaviour of the specialist parasitoid Cotesia plutellae." *Entomol. Exp. Appl., 93,* 87–95.

Ruppert, V & Klingauf, F. (1988). "The attractiveness of some flowering plants for beneficial insects as exemplified by Syrphinae (Diptera: Syrphidae)." *Mitteilungen der Deutschen Gesellschaft für Allgemeine und Angewandte Entomologie, 6*(1-3), 255–261.

Song, B., et al. (2010). "Effects of intercropping with aromatic plants on the diversity and structure of an arthropod community in a pear orchard." *BioControl, 55,* 741–751.

Tentelier, C. & Fauvergue, X. (2007). "Herbivore-induced plant volatiles as cues for habitat assessment by a foraging parasitoid." *J. Anim. Ecol., 76,* 1–8.

Thomas, M. B., Wratten, S. D., & Sotherton, N. W. (1991). "Creation of island habitats in farmland to manipulate populations of beneficial arthropods: predator densities and species composition." *Journal of Applied Ecology, 28,* 906–917.

Tukahirwa, E. M. & Coaker, T. H. (1982). "Effect of mixed cropping on some insect pests of brassicas; reduced Brevicoryne brassicae infestations and influences on epigeal predators and the disturbance of oviposition behavior in Delia brassicae." *Entomol. Exp. Appl., 32,* 129-140.

Van den Bosch, R. & Telford, A. D. (1964). "Environmental modification and biological control." *Biological Control of Pests and Weeds* (P. DeBac, ed.), 459–488. Reinhold.

Wäckers, F. L., Romesis, J., & van Rijn, P. (2007). "Nectar and pollen-feeding by insect herbivores and implications for tri-trophic interactions." *Annual Review of Entomology, 52,* 301–323.

Wäckers, F. L., & van Rijn, P. C. (2012). "Pick and mix: Selecting flowering plants to meet the requirements of target biological control insects." *Biodivers. Insect Pests, 9,* 139–165.

Wade, M. R., & Wratten, S. D. (2007). "Excised or attached inflorescences? Methodological effects on parasitoid wasp longevity." *Biological Control, 40,* 347–354.

White, A. J., Wratten, S. D., Berry, N.A., & Weigmann, U. (1995). "Habitat manipulation to enhance biological control of Brassica pests by hover flies (Diptera: Syrphidae)." *J. Econ. Entomol. 88,* 1171–1176.

Wilby, A., & Thomas, M. B. (2002). "Natural enemy diversity and pest control: Patterns of pest emergence with agricultural intensification." *Ecology Letters. 2002, 5,* 353–360.

Zhao, J. Z., Ayers, G. S., Grafius, E. J., & Stehr, F. W. (1992). "Effects of neighboring nectar-producing plants on populations of pest Lepidoptera and their parasitoids in broccoli plantings." *Great Lakes Entomologist, 24,* 253–258.

Chapter 8: Pollination

Asare, E. (2013). "The economic impacts of bee pollination on the profitability of the lowbush blueberry industry in Maine." Master's thesis, University of Maine, Orono.

Blaauw, B. R., & Isaacs, R. (2014). "Flower plantings increase wild bee abundance and the pollination services provided to a pollination-dependent crop." *J. Appl. Ecol. 51,* 890–898.

Blitzer, E. J., Gibbs, J., Park, M. G., & Danforth, B. N. (2016). "Pollination services for apple are dependent on diverse wild bee communities." *Agriculture, Ecosystems & Environment, 221,* 1–7.

Brosi, B. J., Armsworth P. R., & Daily G. C. (2008). "Optimal design of agricultural landscapes for pollination services." *Conserv. Lett., 1*, 27–36.

Campbell, A. J., Biesmeijer, J. C., Varma, V., Wäckers, F. L. (2012). "Realising multiple ecosystem services based on the response of three beneficial insect groups to floral traits and trait diversity." *Basic Appl. Ecol., 13*, 363–370.

Clark, A. (ed.). (2007). *Managing Cover Crops Profitably*, 3rd ed. Sustainable Agriculture Network.

Gardner, K. E., & Ascher, J. S. (2006). "Notes on the native bee pollinators in New York apple orchards." *Entomologica Americana, 114*(1), 86–91.

Garibaldi, L. A., et al. (2013). "Wild pollinators enhance fruit set of crops regardless of honey bee abundance." *Science 339*, 1608–1611.

Lowenstein, D. M., Matteson, K. C., & Minor, E. S. (2015). "Diversity of wild bees supports pollination services in an urbanized landscape." *Oecologia, 179*(3), 811–821.

MacIvor, J. S. & Packer, L. (2015). "'Bee hotels' as tools for native pollinator conservation: a premature verdict?" *PloS one, 10*(3).

Mader, E., Shepherd, M., Vaughn, M., Black, S. H., & LeBuhn, G. (2011). *Attracting Native Pollinators: The Xerces Society Guide protecting North America's bees and butterflies*. Storey Publishing.

Park, M., et al. (2015). *Wild Pollinators of Eastern Apple Orchards and How to Conserve Them.*, 2nd ed. Cornell University, Penn State University, and The Xerces Society. (www.northeastipm.org/park2012)

Pereira, A. L. C., Taques, T. C., Valim, J. O. S., Madureira, A. P., & Campos, W. G. (2015). "The management of bee communities by intercropping with flowering basil (Ocimum basilicum) enhances pollination and yield of bell pepper (Capsicum annuum)." *J. Insect Conserv., 19*, 479–486.

Shuler, R. E., Roulston, T. A. H., & Farris, G. E. (2005). "Farming practices influence wild pollinator populations on squash and pumpkin." *Journal of Economic Entomology, 98*(3), 790–795.

Venturini, E. M., Drummond, F. A., Hoshide, A. K., Dibble, A. C., & Stack, L. B. (2017). "Pollination Reservoirs in Lowbush Blueberry (Ericales: Ericaceae)." *Journal of Economic Entomology, 110*(2), 333–346.

Winfree, R. W., Fox, J., Williams, N. M., Reilly, J. R., & Cariveau, D. P. (2015). "Abundance of common species, not species richness, drives delivery of a real-world ecosystem service." *Ecology Letters,* 18(7), 626–635.

Index

Page numbers in *italic* indicate photographs.

D

E

F

G

H

I

J

K

L

M

N

O

P

Q

R

S

ADDITIONAL INTERIOR PHOTOGRAPHY BY © agrarmotive/stock.adobe.com, 149; © Alexander Vinokurov/Alamy Stock Photo, 32 r.; © Alexey Protasov/stock.adobe.com, 158; © alexstepanov/stock.adobe .com, 23; © amomentintime/Alamy Stock Photo, 46; © asfloro/stock.adobe.com, 180; © Botany vision/ Alamy Stock Photo, 78 l.; © Brett/stock.adobe.com, 159; © Bryan E. Reynolds, 50, 119 b., 169 l. & c.; © Ciungara/iStock.com, 160; © Engdao/stock.adobe.com, 76 r.; Courtesy Ethan Joseph Smith, 168; © Floki/stock .adobe.com, 95 r.; © fotolinchen/iStock.com, 51; © GabiWolf/stock.adobe.com, 182; © Gillian Pullinger/ Alamy Stock Photo, 111; © GOLFX/Shutterstock.com, 34 r.; Courtesy of Gwendolyn Ellen, Agricultural Biodiversity Consulting, 161; © hhelene/stock.adobe.com, 103; Howard F. Schwartz, Colorado State University, Bugwood.org, 107 l., 140 l.; © hubb67/stock.adobe.com, 82 r.; © imageBROKER/Alamy Stock Photo, 125 b.; © Iva/stock.adobe.com, 80 l.; © jbosvert/stock.adobe.com, 176; © Jessica Walliser, ii, x (ex. t.l.), 2 r., 9, 11, 15, 16, 25, 28, 29, 31, 41, 59 l., 78 r., 83 r., 84, 94, 104 r., 107 r., 120 t., 141, 146, 147, 151, 162–164, 169 r., 170, 171, 178, 179, 181; © John Richmond/Alamy Stock Photo, 82 l.; © Juli/stock.adobe.com, 104 l.; © KaliAntye/Shutterstock.com, 140 r.; © Kelsey/stock.adobe.com, 8; © korkeng/stock.adobe.com, 131 b.; © Lee A. Washington/Shutterstock.com, 98 r.; © Lertwit Sasipreyajun/Shutterstock.com, 98 l., © ligora/iStock.com, 34 l.; © Lithiumphoto/Shutterstock.com, 135; M.A. Brick, Bugwood.org, 37; © M. Schuppich/stock.adobe .com, 125 m.; © Marc/stock.adobe.com, 65; Mary Ann Hansen, Virginia Polytechnic Institute and State University, Bugwood.org, 122 l.; © Nigel Cattlin/Alamy Stock Photo, 17, 102, 136, 138; © nomasa/iStock.com, 44; © ohenze/stock.adobe.com, 18; © Olgalele/stock.adobe.com, 80 r.; © Pernilla Bergdahl/GAP Photos, 12; © photo_pw/stock.adobe.com, 172; © Pravruti/Shutterstock.com, 77; R.J. Reynolds Tobacco Company, R.J. Reynolds Tobacco Company, Bugwood.org, 120 b.; Robert L. Anderson, USDA Forest Service, Bugwood.org, 21; Russ Ottens, University of Georgia, Bugwood.org, 24; © Saxon Holt, 13, 35, 124, 157, 183; Scott Bauer, USDA Agricultural Research Service, Bugwood.org, 109; © sever180/stock.adobe.com, 45; © sharky1/stock .adobe.com, 118 b.; © SJA Photo/Alamy Stock Photo, 20; © Stephen Bonk/stock.adobe.com, 150; © Stephen Studd - Designer: Jon Wheatley/GAP Photos, 14; © tamu/stock.adobe.com, 139; © Tar/stock.adobe.com, 57 r.; © Tim Gainey/Alamy Stock Photo, 148; USDA ARS/Wikimedia Commons, 175; © Varaporn_Chaisin/ iStock.com, 63; © visuals-and-concepts/stock.adobe.com, 125 t.; Whitney Cranshaw, Colorado State University, Bugwood.org, 95 l.

CULTIVATE GARDENING SUCCESS
with More Books from Storey

by Barbara Pleasant & Deborah L. Martin
This thorough, informative tour of materials and innovative techniques helps you turn an average vegetable plot into a rich incubator of healthy produce and an average flower bed into a rich tapestry of bountiful blooms all season long.

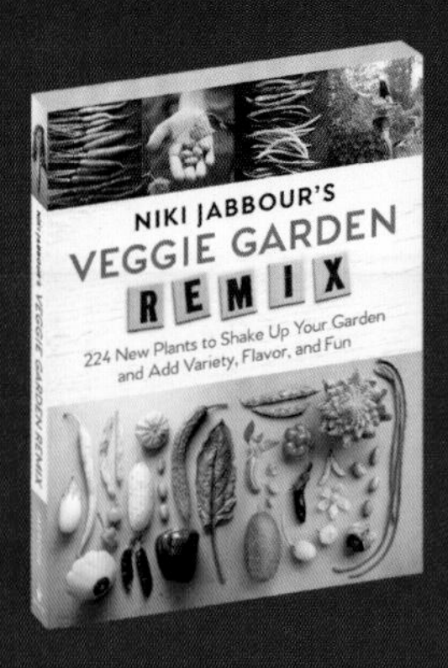

by Niki Jabbour
A lively "Like this? Then try this!" approach starts with what you know, then helps you expand your gardening repertoire by suggesting related varieties and offering detailed growing information for more than 200 plants from around the globe.

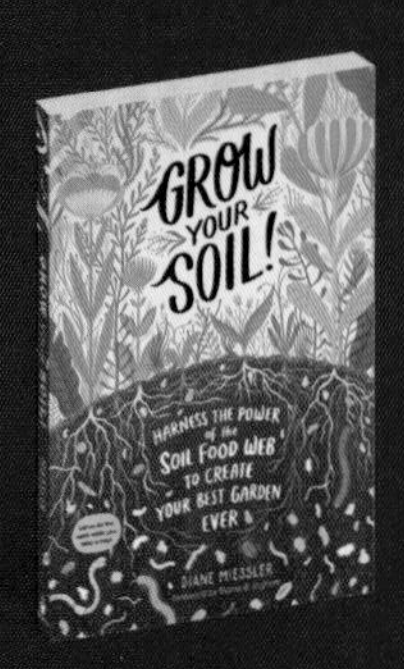

by Diane Miessler
Let microbes do the heavy lifting in your garden! This friendly, eminently readable guide breaks down the science of creating healthy, nutrient-rich soil, with easy-to-follow steps for composting, weeding and mulching strategies, and no-till soil aeration.

by Ron Kujawski & Jennifer Kujawski
These detailed, customizable to-do lists break gardening into manageable tasks. Whether you're planting strawberries, pinching off pumpkin blossoms, or checking for tomato hornworms, this invaluable resource shows exactly what to do — and when and how to do it.

Join the conversation. Share your experience with this book, learn more about Storey Publishing's authors, and read original essays and book excerpts at storey.com. Look for our books wherever quality books are sold or call 800-441-5700.